STORMING HITLER'S RHINE

"A vivid account of the Allied advance on the Rhine in early 1945."—*Booklist*

"Breuer is a first-class historian, and he has written a splendid narrative... factual, balanced and gripping."—*UPI*

"This absorbing account...takes the reader into the war zone to experience the battle as it happened."
—*Concise Book Reviews*

FIRST TIME IN PAPERBACK

STORMING HITLER'S RHINE

THE ALLIED ASSAULT: FEBRUARY–MARCH 1945

WILLIAM B. BREUER

St. Martin's
Press

STORMING HITLER'S RHINE

Printed in the United States of America

First St. Martin's Press mass market edition/August 1986

ISBN: 0-312-90335-9
Can. ISBN: 0-312-90336-7

10 9 8 7 6 5 4 3 2 1

Dedicated to
General J. Lawton "Lightning Joe" Collins,
wartime leader of VII Corps
and later U.S. Army Chief of Staff.
It was an honor
for the author to have served in combat
under this gallant fighting man
in five campaigns in Europe.

No one can last forever. We can't, the other side can't. It's merely a question of who can stand it longer.

If the other side says some day: "We've had enough of it," nothing happens to him. If America says "We're off, we've got no more men for Europe," nothing happens; New York would still be New York, Chicago would still be Chicago, Detroit would still be Detroit. It doesn't change a thing.

But if we were to say today "We've had enough," Germany would cease to exist. That is why we must fight on.

—Adolf Hitler
January 1945

Contents

Introduction   xiii

Part One: Hell-Bent for the Bridges

1. A Dispute Among Generals   3
2. Hitler's Looming Wonder Weapons   17
3. An Angry Patton Threatens to Resign   26
4. "Sleeper Patrols" Among the Enemy   35
5. Assault Over the Roer   48
6. Hitler Rejects a Pullback Over the Rhine   59
7. Vaulting the Erft Canal   71
8. Duel at Adolf Hitler Bridge   81
9. Bloody Road to the Wesel Bridges   94
10. "Mighty Cologne Has Fallen"   106
11. "Your Name Will Go Down in Glory!"   118

Part Two: Over the River

12. "Suicide Mission" at Ludendorff Bridge   133
13. Hitler Thirsts for Scapegoats   146
14. "Annihilation of Two Armies Is Imminent"   157
15. Patton: "Holding the Hun by the Nose"   170
16. "For God's Sake, Tell Them I'm Across!"   180
17. A Daring Plan to Seize Berlin   192
18. "Two If by Sea"   205
19. Crossing the Moat by Buffalo   217
20. The Paratroopers Strike   227
21. Dropping Onto a Hornets' Nest   238
22. The C-46 Deathtraps   247
23. Glider Fighters Pounce   259
24. "General, the German Is Whipped!"   272

Epilogue   283
Notes and Sources   290
Bibliography   297
Index   301

Maps

Allied Lines at the Start of the Rhine Campaign   xi
The Roer River Dams   21
VII Corps' Attack on Cologne and the Rhine   74
Destroying Germans West of the Rhine   109
The Bridge at Remagen   127
The Saar-Palatinate Triangle   164
Operation Varsity Assault Area   215
Operation Varsity Troop Carrier Routes   233
The Rhine Barrier Breached   282

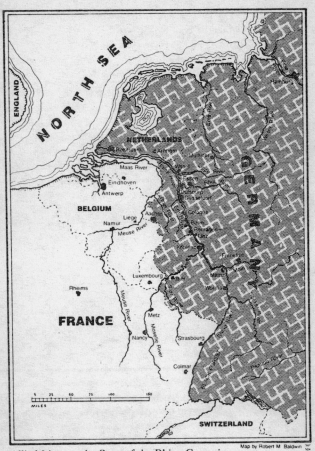

Allied Lines at the Start of the Rhine Campaign

Map by Robert M. Baldwin

Introduction

In early February 1945, with Nazi Germany hemmed in on all sides by hostile forces and on the brink of collapse, Winston S. Churchill placed the Allied situation into perspective: "What is needed is one more gigantic heave!"

Allied Supreme Commander Dwight D. Eisenhower, a former Kansas farm boy, was about ready to launch that final gigantic push. American, British, Canadian, and French armies were stretched out for 450 miles along the western border of the Third Reich from Holland to Switzerland, poised for the signal to fight their way over the Rhine River and into the heart of Adolf Hitler's Third Reich.

The majestic Rhine had been significant in German history ever since Julius Caesar built a timber bridge across it more than 2,000 years before. For 400 years the Rhine was the boundary between the Romans on the west bank and the German tribes. On the west bank grew up the Roman cities of Cologne (*Colonia Agrippina* to the Romans), Koblenz (*Confluentes*), Bonn (*Bonna*), and Mainz (*Maguntiacum*).

During the Middle Ages the Rhine came under German rule from Switzerland to the Netherlands. But when France gained a foothold on the Rhine's western bank in 1648, at the close of the Thirty Years' War, a struggle began between France and Germany for control of the river. This struggle lasted into the 1900s. Louis XIV staked out claims along the Rhine, and Napoleon restored the old Roman boundaries of France. Even after Napoleon was defeated, Alsace, which borders the Rhine from Switzerland to north of Strasbourg, was left in French hands. But Germany seized Alsace again in 1870; and the region reverted to France once more after World War I.

The Rhine originates in eastern Switzerland, flows between Austria and Liechtenstein, and through Germany to the Netherlands, where it empties into the sea, 700 miles from its origin. To Germans, the Rhine was the symbol of their national history and strength. In Richard Wagner's operas, the Nibelungen ring, made from the gold guarded by the Rhine maidens deep in the river's waters, gave its possessor power over all the world.

At Bingen, the Rhine weaves through a deep gorge in the Rhenish Slate Mountains, and there the legendary Lorelei sat on a rock and lured boatmen to destruction with her haunting melodies. Up and down the lengthy waterway rise hills that figure in legends of Atilla the Hun, the heroic Roland, Siegfried, and other historical and mythical figures.

In February 1945, Eisenhower's fighting men had no leisure to contemplate Germany's traditional past. To them, the mighty Rhine was an awesome military obstacle to be surmounted before the war could be concluded. To Adolf Hitler, the Rhine was the final barricade against invasion from the west. No hostile army had crossed it since Napoleon in 1806.

In the looming showdown, General Eisenhower would be holding most of the aces. He had 1,621,027 fighting men arrayed along the approaches to the Rhine, formed into 73 combat divisions—49 infantry, 20 armored, and 4 airborne. Another 2,132,295 Allied soldiers were in service-and-supply roles. Backing this ground force were six tactical commands of fighter-bombers and thousands of medium and heavy bombers. Only three and a half years previously, Dwight Eisenhower had been an obscure staff colonel in Texas who had never led a platoon in combat. Now he was in command of the strongest military force ever assembled.

On paper, the *Wehrmacht* (armed forces) dug in to defend the approaches to the Rhine appeared to be equal on the ground to the Anglo-American armies. Field Marshal Karl Rudolf Gerd von Rundstedt, the 70-year-old aristrocrat known in the Reich as "The Last of the Prussians," was in command of 80 divisions, 7 more than Eisenhower had. But the comparison was illusory. The Allied divisions were at full strength, splendidly equipped, and backed by masses of big guns that were highly mobile. Nearly all of von Rundstedt's divisions were skeletal in their manning levels, poorly equipped, limited in artillery, and lacking in transport and fuel.

Von Rundstedt could expect few replacements, and these would be 15- and 16-year-old boys and men over 45 recently drafted, handed guns, and sent to the front. After five years of war, the Third Reich, with a population of 80 million, had seen 4 million men killed, wounded, or captured.

While General Eisenhower held in his hands almost unprecedented authority to make military decisions involving the land, sea, and air forces of the Western Allies, Field Marshal von Rundstedt was a figurehead, notwithstanding his lofty title of *Oberbefehlshaber West* (Commander in Chief West). Hitler was calling the shots, often down to battalion level, from his underground bunker at the Reich Chancellery in Berlin.

It was obvious to most that Nazi Germany had been crushed. In Munich, Ulm, Düsseldorf, Essen, Hamburg, Cologne, in small towns, tiny hamlets, and in the countryside, civilians were disillusioned with the war—cold, hungry, miserable, and aching for peace. Each day German newspapers carried long, black-bordered lists of the latest young soldiers to fall for the Führer. Hundreds of towns had been pounded into piles of gray rubble by thousands of British and American bombers. Only the most fanatic in the Wehrmacht still believed in victory. Yet one German refused to quit—Adolf Hitler— and he was the only one who counted.

Providentially, many months previously Allied leaders had provided the Führer with the ingredients he now so desperately needed for steeling the resolve of the German fighting man and the civilian population. In 1943, in an offhand statement to news reporters, President Franklin D. Roosevelt had declared that the Allies would accept "nothing less than unconditional surrender."

Prime Minister Churchill, who knew nothing in advance of Roosevelt's ultimatum, was aghast. "Now the Hun [Germans] will fight like cornered rats," he confided to aides. General Eisenhower, who also had been kept in the dark, was equally shocked. "If a German has the choice of mounting the gallows or charging fifty bayonets, he might just as well charge the bayonets," the supreme commander said privately.

General George S. Patton, Jr., who led the U.S. Third Army, told a close friend, "Our president is a great politician, but goddamn it, he's never read history. [He] and the people in our government can't understand the Germans. Look at this goddamn fool unconditional surrender shit. If the Hun ever needed anything to put a burr under his saddle, that's it. Now he'll fight like the devil. It will take much longer, cost us more lives and let the Russians take more territory."[1]

As the Western Allies approached the western frontier of Germany late in 1944, efforts were made to persuade President Roosevelt to soften his stance on the unconditional surrender ultimatum to encourage German capitulation. He refused.

Toward the close of 1944, Hitler was presented with an even bigger propaganda bonanza. A thunderbolt known as the Morgenthau Plan suddenly burst over a disintegrating Third Reich, an uncompromising American proposal for dealing with a defeated Germany. The architect of the plan was Secretary of the Treasury Henry Morgenthau, Jr., who was one of Roosevelt's closest advisors.

Newspapers in the United States obtained copies of the top-secret Morgenthau Plan and made its provisions known to the world. The plan proposed that all German heavy industry be destroyed and that mines in the industrial Ruhr and elsewhere be flooded, thereby denying postwar Germany the raw materials needed for producing peacetime goods. Armed guards would be posted around the borders of Germany with orders to shoot anyone who tried to leave the country.

There was a lengthy list of categories of German officials to be summarily executed. In essence, the proposal would reduce the highly industrialized Germany to an agricultural, pastoral entity, its citizens to be kept in a gigantic ghetto by the conquering Allies.

Hitler and Dr. Josef Goebbels, the Nazi minister of propaganda, were exuberant. Over Radio Berlin and in newspapers, Goebbels milked the drastic Morgenthau Plan for all it was worth, repeatedly pointing out for home consumption that the proposal could not be dismissed as a Big Lie from the Third Reich leadership because it emanated directly from Washington, D.C. Germany, Goebbels trumpeted, would be reduced by vengeful Allies to a nation of 80 million serfs, fed from soup kitchens and bread lines until the end of time.

Adolf Hitler's further hope that the Allied alliance would come apart and allow Nazi Germany to gain a negotiated peace gave indications of becoming a reality in early 1945. With the conflict in Europe apparently nearing an end, generals and heads of state were already maneuvering for position in the postwar world. Personality clashes, sparked by powerful egos, nationalistic goals, and festering suspicions, racked the Western Allies. Tempers were short, accusations bitter.

British Field Marshal Alan Brooke, the British Army Chief of Staff, scrawled in his diary: "Ike [Eisenhower] has a very limited brain . . ." He noted that "Eisenhower has only the vaguest grasp of strategy and tactics." Brooke was convinced that he knew more about winning the war than did either Eisenhower or General George C. Marshall, the U.S. Army Chief of Staff. He doubted Marshall's qualifications for his high post and thought that General Douglas MacArthur could do a much better job as chief of staff.

Brooke's assessment of Eisenhower and Marshall must be seen in the context of his own shattered professional aspirations. Churchill had promised the coveted supreme commander's post to Brooke, but it went to Eisenhower—at the insistence of Marshall.

For his part, the 63-year-old, otherwise dignified George Marshall was barely able to conceal his contempt for his British opposite number, Brooke. Marshall was ceaselessly irritated by what he considered Brooke's condescending attitude toward the "newcomers to war" from the colonies across the Atlantic. The American chief of staff held an equally low regard for Field Marshal Bernard L. Montgomery, leader of the 21st Army Group, whom Marshall considered the source of much friction between the Western Allies.

The wiry and assertive Montgomery, a living legend on the British homefront and the darling of London's Fleet Street press, regularly got in his licks against Supreme Commander Eisenhower. In late 1944, he had written his boss and mentor, Alan Brooke: "If you want to end the war in any reasonable time, you will have to remove Ike's hand from control of the land battle."

Forbearing and politic, Eisenhower had finally grown weary with what he considered relentless British efforts to undermine his role and authority. "If they [the British] keep this up, then I'll go home!" he exploded to confidants. The supreme commander even drafted a letter to General Marshall with an ultimatum: "Either Montgomery goes or I go." Cooler heads prevailed, and that potentially disastrous move was averted.

Neither Lieutenant General Omar N. Bradley, the soft-spoken former Missouri school teacher who commanded the 12th Army Group, nor General Patton was on comfortable speaking terms with Field Marshal Montgomery. Patton privately referred to Montgomery as "that little fart," and Bradley regarded the highly successful British commander as "an arrogant egomaniac" and a "third-rate general who never did anything that any other general could not have done better."

When Hitler launched two panzer armies out of the rugged fir forests of the snow-blanketed Ardennes of Belgium and Luxembourg on December 16, 1944, he caught the Americans by total surprise—"with our pants down," as an Eisenhower staff officer phrased it privately. In what came to be known as the Battle of the Bulge, Bradley's army group was split in two by German spearheads plunging into Belgium, making it difficult for Bradley to communicate with his First Army north of the enemy penetration. As a result, Eisenhower turned over command of American armies on the north flank to Montgomery, leaving a chagrined Omar Bradley with only Patton's Third Army.

Bradley was furious over losing half of his army group. His anger turned to raging apoplexy when, after the Bulge battle was won, Montgomery called a press conference on January 7 and told reporters how he had "tidied up the mess"—a "mess" for which Bradley was presumably responsible. Headlines such as the one that appeared the next day in the *Times* of London poured oil on the fires of American-British acrimony:

> MONTY FORESAW DANGER;
> TOOK STEPS ON HIS OWN
> TO HEAD OFF DISASTER

Eisenhower, Bradley, and Patton—especially Bradley—were hopping mad. If the British field marshal had "foreseen danger," they thought, he had a peculiar reaction to it. On December 15—one day before Hitler struck in the Ardennes—Montgomery had asked for and received permission from Eisenhower to "hop over to England to spend Christmas with my son." Omar Bradley would never forgive Bernard Montgomery for his press conference remarks.

Now, in early February 1945, as General Eisenhower prepared to send his armies storming across Germany's historic river barrier against invasion from the west, the Allied high command was racked with dissension. American and British generals were eyeing each other suspiciously, like two dogs going after the same bone. Damage was still evident from the verbal salvos that had been fired by both sides, and there were machinations on both sides to be first over the Rhine.

As tension mounted along both banks of Germany's wide moat, Adolf Hitler, in far-off Berlin, was determined that both Americans and British would fail. He intended for the Rhine to run red with Allied blood.

I

Hell-Bent
for the
Bridges

1

A Dispute Among Generals

Panic and chaos gripped a drab gray Berlin in early February 1945. The once majestic city was now a wasteland of fallen masonry. Milling, ragged, half-frozen hordes of refugees stumbled into the Nazi capital from the east, gasping out tales of mass looting, murder of civilians, and wholesale rape by the Russian army now only 90 miles away. Each night Berliners, risking the death penalty, tuned clandestine radio sets to the BBC in London and were made aware that powerful Anglo-American armies were massing across the Rhine River to the west.

Berlin had not been captured by a foreign invader since 1806—and there had been no battle then. Headed by their mayor, the citizenry had emerged to welcome Napoleon's army, and the German press had lavished praise on the French and their occupying soldiers. Now, 139 years later, Berlin citizens were, in effect, part of the German garrison, joining the desperate effort to defend the capital against the impending arrival of Anglo-American forces from the west and the Red Army from the east.

The *Volkssturm* (people's army) of young boys and old men who felled trees, dug trenches, and fashioned barricades in and around the outskirts were simply civilians with red arm bands. Women rode anti-aircraft guns pulling out of Berlin toward the Russian front, many never to be heard from again. The "mayor" this time was Dr. Josef Goebbels; the propaganda minister had been given the additional title of *Gauleiter* of Berlin.

Over Radio Berlin, the diminutive, 47-year-old Goebbels screamed defiance: "Berlin and all of the Fatherland will be defended stone by stone!" Robert Ley, Hitler's labor chief, took to the airwaves to exhort the Wehrmacht and the German people: "We will fight in front of Berlin, in Berlin, behind Berlin!"

Pouring out of Radio Berlin night and day was a constant stream of exhortations to "true" Germans to resist the invaders, mingled with threats to "traitors, shirkers, and defeatists." All males of the city over 13 years of age were ordered to report immediately for Volkssturm duty. Busses were halted, refugees handed guns and sent to the fighting fronts, much in the manner of Marshal Gallieni's French army, which was rushed to the ramparts in a fleet of taxis to help save Paris from the Germans in 1914. Food cards were stretched out for an additional week, and rations were cut—again.

Civilian automobiles had all but disappeared from Unter den Linden, Potsdamer Platz, and other Berlin streets. Trams ran sporadically. Someone broke into a radio broadcast shouting "*Schluss! Schluss!*" (This is the end!), followed by scuffling sounds, a brief silence, and then the blaring of patriotic marches.

Rumors were rampant. A huge, ultrasecret new airplane was standing by to fly Nazi bigwigs to Japan, said one report. Hitler, Reichsmarschall Hermann Göring, Goebbels, Reichsführer Heinrich Himmler (chief of the dreaded Gestapo and commander of the elite SS), and other Nazi functionaries were said to be carrying vials of poison in the event Anglo-American or Russian paratroopers made a sudden strike to seize the German leaders.

Always there was the eerie wail of air-raid sirens. Big Halifax and Lancaster bombers of the Royal Air Force came by night, 1,200 of them at a time in a stream stretching back for a hundred miles. They showered Berlin with thousands of high-explosive and incendiary

4

bombs, obliterating huge areas and reducing rubble piles still further into powder. Buried under tons of debris after each raid would be hundreds of Berliners—dead, half-alive, mutilated. A short respite, then with daylight the sirens would moan again, a warning that a teeming conveyor belt of American Flying Fortresses and Liberators was beginning again.

Germany was, as Franklin D. Roosevelt had observed, a fortress without a roof. The air war put everyone in the Third Reich in the front lines—including the Führer. The gardens of the Reich Chancellery in Berlin, where Hitler was directing the siege of Germany, were pockmarked by deep craters. Fallen trees and piles of rubble barricaded the garden paths where Hitler had loved to stroll and contemplate his dreams of world domination. The Führer's ornate residence nearby had been struck several times by bombs, and only the unsteady walls remained of the huge banquet hall where the founder of National Socialism had so often harangued the faithful.

Light, power, and water were missing at the seat of government in dying Berlin. A water-tank wagon was parked in front of the Reich Chancellery, and soldiers periodically dashed outside to haul in buckets of water for cooking and washing.

Martin Bormann, the squat, moon-faced Nazi Party functionary who had worked his way up the totem pole of influence to become Hitler's most trusted confidant, early in February complained of conditions in a letter to his wife: "The worst thing of all is the water closets [toilets] in the Reich Chancellery. These Kommando pigs [military officers] use them constantly, and not one of them ever thinks of taking a bucket of water to flush the place . . ."[1]

Meanwhile, on February 3, hundreds of miles from Berlin, at Yalta in the Crimea, the Big Three powers were convening in a summit conference of U.S. President Franklin Roosevelt, British Prime Minister Winston S. Churchill, and Russian Premier Joseph Stalin. The purpose: to apportion the postwar world.

President Roosevelt, his face etched with pain, was a dying man. Before departing Washington, Roosevelt, a shadow of his former robust self, called in a trusted advisor, the white-haired, courtly Bernard Baruch, for advice on the forthcoming session of the Big Three, where historic arrangements would be hammered out. Seated in the White House oval office, Baruch was shocked by the president's haggard appearance. He noticed that Roosevelt's hands were trembling. By way of explanation, the president gave his famous lopsided grin and said breezily, "Bernie, I had too many with the boys last night."[2]

Baruch had no way of knowing that the ailing president would be dead in nine weeks, 83 days after he had been sworn into office for an unprecedented fourth term. The dignified Baruch told the president, "The Bible and history are filled with missions upon which countless men have set forth to help their fellow men. Never has one been fraught with such possibilities as the one upon which you are about to embark."

Inspirational words, these. A more practical-minded aide later cautioned Roosevelt to "keep your guard up" when dealing with Stalin. Again the smile. "Don't worry about it," the president responded airily. "I can handle Uncle Joe."

"Uncle Joe" would prove at Yalta to be more than a match for the exhausted, deteriorating president and the aging, war-weary Churchill. Outwardly affable, yet cunning, ruthless, and mentally tough, the five-foot-six Stalin had little difficulty in mesmerizing his two allies into giving him Poland, Czechoslovakia, Hungary, Austria, Bulgaria, Rumania, and the Baltic countries.

The groundwork was laid for slicing up a vanquished Germany into three parts, or possibly four if France was accepted as a full Allied partner. Roosevelt had been deeply impressed by the devastation he had seen in the Crimea while driving to the conference, and told Stalin he was "more bloodthirsty" than ever regarding the Germans. At an evening social function, Roosevelt remarked to Stalin, "I hope you'll again propose a toast to the execution of fifty thousand officers of the German army."[3]

While the Big Three divvied up postwar Europe, a short distance away the Anglo-American Combined Chiefs of Staff, the panel charged with direction of the war in the west, had convened to hear Supreme Commander Eisenhower's plan for getting across the Rhine and advancing into the heartland of Germany. Almost immediately the most heated controversy of the war erupted.

At this time, the 450-mile Allied front had Montgomery's 21st Army Group, consisting of three armies—the Canadian First, the British Second, and the U.S. Ninth—on the north. In the center was Bradley's 12th Army Group, comprised of the U.S. First and Third armies. In the south, led by Lieutenant General Jacob L. Evers, was the 6th Army Group, which had the U.S. Seventh and French First armies.

As American and British brass listened intently, Lieutenant General Walter B. "Beetle" Smith, Eisenhower's sharp-tongued chief of staff, outlined the plan drawn up at SHAEF (Supreme Headquarters, Allied Expeditionary Force). He stressed that timing of the "one more gigantic heave" was of the utmost significance, and that the time to strike was now, while the embattled Wehrmacht was trying desperately to hold off Russian attacks in the east. On January 12, nearly 3 million Red Army soldiers, backed by T-34 and Stalin tanks, had overrun some 750,000 ragged and frozen German soldiers along a 390-mile front. The Russians had plunged ahead for 200 miles and were now 90 miles from their ultimate objective, Berlin.

Eisenhower's strategy was first to push all Allied forces up to the Rhine. Then Montgomery's 21st Army Group would launch the major assault over the water barrier in the northern sector. The SHAEF decision to give the ball to Montgomery was based on a logical military reality, Smith stressed: once over the Rhine, Monty had by far the most favorable terrain north of the Ruhr Basin industrial region, over which his forces could advance rapidly.

Montgomery's Rhine-crossing operation would be enormous, comparable to Neptune, the D-Day assault against Normandy the previous June. Supporting his river vault would be thousands of heavy bombers and fighter-bombers, parachute drops and glider landings by at least two airborne divisions, an unprecedented mass of artillery and mortars, and a major naval amphibious operation involving hundreds of assorted craft.

Eisenhower's plan appeared to be a victory for the British strategy. Led by the persistent lobbying of Field Marshals Brooke and Montgomery, the British for weeks had been arguing for a single powerful thrust into the Third Reich heartland with Montgomery in charge. American generals had opposed the British proposal of a single thrust, which would result in powerful American armies sitting idle along 300 miles of the Rhine River's west bank while Montgomery raced to Berlin.

But the apparent British victory at the conference table was not to be total. Despite the gigantic massing of troops, firepower, and air support in Montgomery's crossing, there was always the possibility of failure, General Smith stressed. Therefore, Omar Bradley's 12th Army Group would make a *secondary* attack. Eisenhower, a consummate statesman, cited sound military logic for his proposed double-pronged attack.

The strategic compromise left both Americans and British unhappy—each suspicious of the supreme commander's motivations. Smith had no sooner concluded his presentation and been seated than Field Marshal Brooke jumped to his feet and attacked the Eisenhower plan. He especially assaulted the proposal for clearing the entire west bank of the Rhine before Montgomery leaped across the water barrier. General George Marshall, the U.S. chief of staff, listened in silence, his neck growing red with anger. When Brooke concluded, Marshall lashed out in defense of Eisenhower. The dispute that followed became so heated that Marshall demanded a closed session, with only the principals remaining in the room and the note-taking stenographers dismissed. Velvet gloves were replaced with verbal brass knuckles, diplomatic niceties flew out the window, and strategic plans to administer the coup de grace to Nazi Germany were temporarily forgotten as a scathing denunciation of personalities erupted. Brooke shouted that the British chiefs were deeply concerned because Eisenhower was "overly influenced" by Bradley.

"Well, Brooke," snapped Marshall, "they are not nearly as much worried as the American chiefs are about the pressures of Mr. Churchill on General Eisenhower."[4]

Marshall pointed out heatedly that at his (Marshall's) suggestion, President Roosevelt almost never saw General Eisenhower, "because Eisenhower was not just an American but an Allied commander." He declared that he was "deeply concerned" by the constant pressures brought against Eisenhower by Churchill and the British chiefs of staff.

General Marshall then lit into Field Marshal Montgomery. He exclaimed that Brooke himself was "overly influenced" by Monty, and that much of Brooke's distrust of Eisenhower and his plan for delivering the knockout blow to Nazi Germany resulted from Montgomery's machinations. The stern-faced British chiefs, gripping chair arms in white-knuckled fury, heard Marshall bluntly assert that he wanted to "express my full dislike and antipathy for Field Marshal Montgomery."[5]

For months, with the backing of Alan Brooke, Montgomery had been conducting a barely disguised campaign to have a "land commander" appointed over all Anglo-American forces. (There were already sea and air commanders under Eisenhower.) Montgomery's candidate for the post: Bernard Montgomery. Since American and British troops had secured a firm foothold in France on September 1, General Eisenhower had been performing in the dual role of land commander and supreme commander.

Now at Malta General Marshall brought the simmering issue into the open. Previously he had told Eisenhower that he would resign as chief of staff if Eisenhower were "saddled" with Montgomery as land commander. Now he made it forcefully clear to the British chiefs that he would not accept Montgomery "or anyone else" in that post.

As Marshall concluded, a thick pall of silence enveloped the conference table. Brooke launched a tactical retreat. He declared that if Eisenhower would give him assurances "in writing" that "every single German soldier" would not have to be cleared from the west bank of the Rhine before Montgomery leaped the water barrier, and that Bradley's secondary effort would not be converted into a major assault, Brooke would give his approval to the Eisenhower plan.

Field Marshal Brooke's unspoken concern was that the secondary thrust might become as significant as Montgomery's main effort. The British chief felt justified in his belief that any attack involving General George Patton was bound to develop into a full-blooded offensive—whether it was planned that way or not.

Silver-haired George Marshall had scored a resounding double victory at the conference table. He had contained British demands for Montgomery to be appointed land commander and ensured that Bradley's powerful First and Third armies would not languish at the rear while Field Marshal Montgomery made a triumphant entry into Berlin.

Far from the heated disputes at Malta, along the fighting front west of the Rhine River, a grumpy George Patton had just returned to his headquarters in Luxembourg from a round of visits to his units. Patton was in a foul mood. His Third Army was hacking its way forward, but the rugged terrain—hilly, wooded, cut by deep ravines—had made the going slow and costly.

Patton had just lighted a cigar when Omar Bradley telephoned. "Monty did it again, George," the 12th Army Group leader declared. "You and Hodges [First Army] will go on the defensive while Montgomery will resume the offensive in the north." There was a brief silence, then Bradley continued: "It wasn't Ike this time. Orders from the combined chiefs. Even Marshall went along. . . . Probably he's anxious to get those fourteen British divisions sitting on their asses in Belgium back into action."[6]

"What in the goddamned hell are they hoping to accomplish?" Patton exploded.

"Montgomery wants to secure a wide stretch of the Rhine in case Germany would suddenly collapse."

"Horse shit!" Patton retorted. "We've a much better chance to get to the Rhine first with our present attack." Then he added: "What in the hell are we supposed to do in the meantime?"

"You can continue your attack until February tenth, and maybe even after that," Bradley replied, "provided your casualties are not excessive and you have enough ammunition."[7]

That night, George Patton vented his anger in his diary. In his nearly illegible handwriting and faulty spelling he scrawled: "Hell and damn! This is another case of giving up a going attack in order to start one that has no promise except to exalt Monty, who has never won a battle since he left Africa (and I won the Mareth Line for him there) . . ."

Despite the protests of Bradley, Patton, and Lieutenant General Courtney H. Hodges of the First Army, the big troop shift to Montgomery got underway. Monty already had Lieutenant General William H. Simpson's Ninth Army for the Rhine crossing assault, and now the 95th Infantry Division and five artillery battalions were taken from the Third Army and given to Simpson. Patton, bitter and frustrated, wrote in his diary: "I will be the first to the Rhine yet!"

None of the American battle generals faulted Simpson for accepting the "stolen" divisions from Bradley's army group. A tall, energetic, and keen-witted infantryman, "Big Simp" was liked by all. He possessed an engaging sense of humor and often joked about his clean-shaven head. Once in training camp in the United States he had issued a tongue-in-cheek directive which asserted that all officers would get haircuts "precisely like that of the commanding general."

Simpson's Ninth Army, which had been bloodied in previous months while fighting to wrest several ports in Brittany from die-hard German troops, had been assigned to Montgomery's British army group by the happenstance of being on his right flank. But it was an ideal inter-Allied marriage. Simpson, as affable as he was able, overlooked what other American commanders considered Montgomery's arrogance and meshed harmoniously with the idiosyncratic field marshal and his staff.

During these days of sharp infighting between American and British generals, Supreme Commander Eisenhower was taxed by his role as referee of the disputes. In his delicate post as leader of an international coalition, Eisenhower often had to say no to all sides. It was not long before Omar Bradley telephoned the supreme commander in a fit of frustration and demanded that Simpson's Ninth Army be returned to him.

"The public back home is sore as hell about Monty's continuous publicity," Bradley declared. And he pointed to a fact about which

Eisenhower was already aware: the United States had 61 of the 73 Allied divisions arrayed along the western border of Germany.[8]

Eisenhower refused his 12th Army Group commander's request, patiently sketching once more the strategic reasons for assigning Big Simp's Ninth Army to Montgomery. But the supreme commander, whose temper had a short fuse and who could unlimber a muleskinner's vocabulary when angered, was himself furious at Montgomery for his remarks to the British press in recent days while the field marshal was on a short leave in London. He called his British aide, Brigadier John F. M. Whitely, into his office and ordered him to telephone Montgomery in England. "Tell him if one member of the Twenty-first Army Group talks to the press," Eisenhower told Whitely with considerable heat, "I'll turn over command [of the Rhine push] to Bradley."[9]

Whitely blanched. He was in the unenviable position of having to telephone a senior fellow British general, several ranks his superior, and warn him, in effect, to keep his mouth shut. Whitely made the call.

That night, Lieutenant Kay Summersby, the British subject who had been commissioned in the United States Army and had been an Eisenhower confidante for more than two years, wrote in her diary: "Of all of E.'s commanders, Monty is the one who has given him the greatest number of headaches."

By early February, the American press had taken a flip-flop from its heady optimism of the previous fall when the Allies had been racing across France and Belgium for the German border almost unopposed. Now the press conveyed gloom and predicted that the Nazis would fight to the last man and find some diabolical method for prolonging the war indefinitely. The best locale for defense was in the Alps, and it was there that the Germans had built a last-stand bastion. The media even gave the stronghold a catchy name: Hitler's Alpine Redoubt. Centering on the picturesque little town of Berchtesgaden, where Hitler had a retreat perched atop a Bavarian mountain, the redoubt would be protected by the towering, inaccessible heights of southern Germany, Austria, and northern Italy.

The Nazi alpine stronghold had been a mammoth undertaking, an engineering marvel, stories pointed out. Factories had been burrowed into mountainsides; enormous amounts of food, medical supplies, and other necessities had been stored in underground shelters. There was a maze of subterranean tunnels and living quarters, gun positions, and ammunition dumps, all impervious to Allied bombing. Finally, when all appeared lost, Hitler and other top Nazis would pull back into this impregnable keep along with 250,000 die-hard SS troops, and defy efforts to dislodge them until America and Great Britain grew weary of the unending struggle and settled for a negotiated piece.

American and British radio commentators got into the act, sometimes letting imaginations run free. Allied intelligence took notice of Hitler's Alpine Redoubt, partly due to the drumfire of civilian newspaper stories and radio broadcasts, and soon was issuing reports on the stronghold. Bits and pieces of evidence culled from many sources provided documentation for the alarming intelligence reports. General Eisenhower and SHAEF became increasingly worried.

Hard on the heels of the Hitler Redoubt revelation, another last-gasp Nazi threat surfaced. Allied intelligence and the civilian press—one feeding upon the other—told of a secret movement being formed within the Third Reich called Werewolves. In mythology, a werewolf was a person capable of transforming himself into a wolf and preying on unsuspecting humans. Trained and organized by Heinrich Himmler's SS, members of the Werewolves were zealous, even fanatic *Hitlerjugend* (Hitler Youth) who had taken blood oaths to die for the Führer if need be. They would remain in territories overrun by the Allies, murder Allied soldiers and Germans cooperating with the enemy, and blow up the invaders' facilities—so the reports declared. The tender years of a German Werewolf, and the fact that he would be operating in his home locale, would make it nearly impossible for the invading Allies to identify him.

Many weeks would pass before the Allies realized that they had been the victims of a monumental hoax perpetrated by Josef Goebbels. The Hitler Alpine Redoubt was a total myth, manufactured by Goebbels's fertile brain to encourage the embattled Wehrmacht and the war-weary German people to fight on in a hopeless situation. Efforts were in fact made to create a Werewolf organization, but they foundered against the enormity of the task and the apathy of most prospective recruits. So intense had been the barrage of Allied media reports and intelligence assessments that Eisenhower, Bradley, and staff officers at SHAEF had accepted the myths as fact. It would have taken years to plan, design, and construct an underground fortress of such scale, yet at each SHAEF strategic planning session the haunting specter of Hitler's Alpine Redoubt hovered over the conference and influenced far-reaching decisions.

In bomb-ravaged Berlin on the afternoon of February 2, Colonel General Heinz Guderian, the red-haired, short-tempered, able army chief of staff and commander of the long Russian front, strode with his aide, Major Baron Bernd Freytag von Loringhoven, through the huge oak door of the Reich Chancellery for a high-level conference of military commanders with Adolf Hitler. Guderian and von Loringhoven walked briskly down lengthy corridors to an anteroom of Hitler's cavernous private office, where they were halted by black-uniformed, stern-faced *Schutzstaffel* (SS) men with Schmeisser machine pistols at the ready. General Guderian and his aide then submitted to an indignity that had been standard practice since the previous July 20 when several members of the German general staff had been involved in a bomb plot to murder Hitler, an assassination effort that had gone awry. Guderian and von Loringhoven were ordered to hand over their personal weapons to the SS guards, and their briefcases and persons were scrupulously searched. Hitler no longer trusted his generals.

A glittering array of Wehrmacht brass filed into the Führer's office to await Hitler's appearance. They were all there, including the increasingly obese Reichsmarshal Hermann Göring, the number two Nazi, who had taken to wearing rouge, lipstick, and fingernail polish at his luxurious mansion in the countryside north of Berlin. Göring, rumor had it among the German generals, had become addicted to narcotics and often rambled in conversation.

Suddenly, a massive door opened at one end of the office and in shuffled Hitler and two aides. The high brass came to their feet as one, clicked heels, and stood at attention until Hitler took his seat at a table. Those who had seen him infrequently were shocked—he was haggard and looked far older than his 55 years. He was stoop-shoul-dered, one arm hung limply at his side, and those seated close to him noticed his hands trembling.

Nearby generals noticed through furtive glances that by careful ma-nipulation Hitler was able to reach under the table with his right hand, move his limp left hand onto the desk, and thereby manipulate documents and maps. There had long been speculation as to the cause of Hitler's useless left arm. Some thought it was a result of the bomb explosion the previous July; others held the opinion that he had Par-kinson's disease, or that he was an epileptic. Due to the secrecy in which his health was inevitably shrouded, it was impossible to know the precise cause of the infirmity. They did know that Hitler relied heavily on Dr. Theodor Morell, his personal physician, to keep up his strength with strychnine injections and Benzedrine pills.

Hitler's two companions as he entered the room, Field Marshal Wilhelm Keitel and Colonel General Alfred Jodl, were long-time con-fidants and among the remaining handful of men Hitler trusted. Kei-tel, humorless and flint-faced, was sort of a Wehrmacht minister of the interior, an administrator at the highest level who did not, in the strictest sense, intervene in military field operations. As *Chef des Oberkommandos der Wehrmacht* (Chief of the High Command), he had been at Hitler's elbow since 1940. But Keitel was held in low regard by German combat generals and was considered to be a toady. He was said to have had an ambition to become a farmer; Wehrmacht generals whispered privately that he had missed his calling. Many pinned him with the derisive moniker "Lakeitel" (from *Lakai*, lackey). For five years Keitel had been one of the most influential men around the Führer and had played a major role in Nazi Germany's expan-sionist policies, which at one time had gained Hitler control of nearly all of Europe.

Colonel General Jodl had held his office, *Chef des Wehrmachtführungsstab* (Chief of the OKW operations staff) since the start of the war in September 1939. He was the man designated by Hitler to translate strategic decisions into precise orders for distribu-tion to commanders in the field. In the clumsy scheme of things in Hitler's conduct of the war, Jodl's responsibility was limited to the western front and the Italian theater. (The Russian front was the domain of General Guderian, known as "Hurryin' Heinz" for his armored dashes earlier in the war.) There was a good reason for this division of responsibility. The paranoiac Hitler did not want too much power concentrated in the hands of any one general.

An eerie hush fell over the room as Hitler eyed the commanders arrayed before him. General Guderian opened with a long and brutally frank analysis of the Russian front. Hurryin' Heinz was one of the few remaining generals who was honest with the Führer when speaking of battlefield situations. Hitler showed little interest in the east, but his attention picked up noticeably when Göring began talking about the situation along the Rhine River in the west.

"Deep down, don't you think the English really are unhappy about the Russian successes?" Hitler interjected. He always referred to the Anglo-American alliance as "the English."

"If this continues," Göring declared sarcastically, "we'll be getting a telegram soon [from the English]. They didn't expect us to defend ourselves step for step and hold them off in the west like lunatics while the Russians drove deeper and deeper into Germany."[10] Like Guderian and most other Nazi generals, Göring considered it absurd to fight so tenaciously against the Anglo-Americans in the west while the eastern front was collapsing.

The conference rambled on for two and a half hours. Hitler interjected anecdotes of his experiences as a frontline infantry corporal in World War I into discussions of the monumental problems facing Germany on the far-flung battle lines and at home. It was nearing 7:00 P.M. when General Guderian and his aide climbed into a staff car for the 20-mile ride back to Guderian's headquarters at Zossen, south of Berlin. The general was livid. Not a single decision for halting the Anglo-American and the Russian tides on the west or east had been reached.

2

Hitler's Looming Wonder Weapons

Bundled up in his salmon-colored trench coat against the damp wind, Supreme Commander Dwight Eisenhower drove up to Omar Bradley's 12th Army Group headquarters at Namur, Belgium, on the night of February 4, 1945. Eisenhower had summoned Field Marshal Montgomery to meet him there the following day. "Ike," as the supreme commander had been called since childhood, and Bradley, two old friends, talked far into the night. Bradley was still upset over Montgomery's January 7 press conference, in which the field marshal had said he had "tidied up the mess"—Bradley's "mess"—in the Battle of the Bulge.

The British field marshal now had 15 American divisions under his command, and Bradley feared that London newspapers would soon ballyhoo how Montgomery once again was coming to the rescue of the Americans, this time in the drive to and over the Rhine River. Eisenhower told the agitated Bradley of how he had directed Brigadier Whitely to call Montgomery and tell him to "keep his mouth shut" on pain of having Bill Simpson's U.S. Ninth Army returned to

Bradley for the Rhine assault. The 12th Army Group commander was only partially placated.

When Bernard Montgomery strode breezily into Bradley's headquarters the next morning, he was greeted by the Americans with a chill equal to the frigid weather outside. If the field marshal noticed the tone of the reception, he gave no indication of it. During lunch, Montgomery joked, gesticulated, and talked constantly—and too loudly, the stone-faced Americans thought.

The luncheon ended quickly. Omar Bradley abruptly got to his feet, excused himself without looking at the visiting British commander, and left Montgomery still seated at the table. Minutes later, Bradley and Eisenhower climbed into the supreme commander's olive-drab Packard Clipper and raced off for a previously planned tour of the front. They headed for Bastogne, where they met General Patton—"Georgie" to his fellow generals. Patton had driven up from his headquarters in Luxembourg and, typically, was brimming with enthusiasm and ideas for breaking through to the Rhine; but Ike put a damper on Patton's euphoria, reminding him that he was to conduct "an aggressive defense" while Montgomery, far to the north, prepared to assault the wide water barrier.

Before leaving Montgomery at Namur earlier in the day, Eisenhower had cautioned the field marshal once more about a repetition of his gratuitous remarks to the press following the Battle of the Bulge. Now it was the volcanic Patton's turn to be warned about a loose lip. Keep a low profile, Eisenhower admonished the flamboyant Third Army commander, or else Montgomery might take umbrage. Old cavalryman Patton's eyes flashed and his face flushed with anger. He started to say something, but held his tongue and made no reply. The supreme commander moved on to another subject. (To a handful of confidants at SHAEF, Eisenhower referred to Montgomery and Patton in periodic fits of exasperation as "my two prima donnas.")

Darkness was descending on the battlegrounds west of the Rhine on the evening of February 5 as Generals Eisenhower and Bradley drove up to the U.S. First Army headquarters at Spa, Belgium, the peacetime resort tucked away in the Ardennes forest. General Courtney Hodges, the reserved, soft-spoken First Army commander, and his staff had been forced to flee for their lives at the last minute seven weeks previously when swarms of tanks of the 1st SS Panzer Division broke into Spa in the opening stages of the Battle of the Bulge. Now Hodges had returned his command post to its former location.

Eisenhower found the normally imperturbable Hodges on edge. First Army had been assigned the crucial mission of capturing seven Roer River dams, some 40 miles west of the major German city of Cologne on the Rhine. Hodges had given the task to Major Edwin P. Parker, Jr.'s, largely inexperienced 78th Infantry Division. Handicapped by the rugged terrain, oceans of mud from the thawing winter snows, heavy forests, and tenacious German resistance, Parker's men were having a tough—and bloody—time hacking their way to the dams. "Your mission is the most vital one at this time on the entire Western Front," Parker had been told by Major General Clarence R. Huebner, commander of V Corps. "These dams have to be taken—and soon!"

Downstream (north) of the dam complex elements of Hodges's First Army and Simpson's Ninth Army had advanced to the normally narrow, serene Roer River, which runs roughly parallel to the Rhine. The floodgates there had been arranged to cause the worst possible conditions downstream if German demolition engineers blew them up. Such an eventuality would result not in a sudden, brief torrential burst of water surging in front of Hodges's and Simpson's forces, but in a steady flow over a prolonged period that would have turned the gentle Roer into a raging torrent and flooded the valley to a width of a mile in some places.

With the advancing Americans threatening to overrun the dams the previous November, Adolf Hitler, from his headquarters hundreds of miles away, had ordered the seven huge dams destroyed. But his commander in the sector, General of Panzer Troops Baron Hasso von Manteuffel, at considerable risk to his well-being, had overruled Hitler. Recognizing the vital role the dams would play in stemming the Allies' headlong charge to the Rhine, von Manteuffel had ordered the Roer dams held at all costs.

One of these, the Erft Dam, had been constructed just after the turn of the century on the Erft River near its confluence with the Roer. It was made of concrete and was over 700 feet long. The Schwammenauel Dam, which held twice as much water as the enormous amount penned up by the Erft Dam, was of earthen construction with a concrete core. Built by German engineers in the mid-1930s, it was located on the Roer River a few miles north of the Erft Dam.

The dams for months had been a frightening bugaboo to American soldiers. From the previous September through early December, General Hodges had launched four different divisions in four separate all-out attacks through an almost trackless forest to seize the dams. Each division had been cut to pieces. The 28th (Keystone) Infantry Division, a Pennsylvania National Guard outfit, had been virtually destroyed the previous November. In seven days the 28th Division had suffered 6,184 casualties, one of the heaviest divisional bloodlettings in the American army during the war.

Royal Air Force torpedo bombers were brought in to smash the dams and release the flow of pent-up water, but the torpedos did as little damage to the massive Schwammenauel Dam as a flea would to the hide of a rhinoceros.

Pressure to rapidly seize the key Roer dams was being exerted throughout the Allied chain of command. Grand strategy for the advance to the Rhine called for the dams to be in American hands before Field Marshal Montgomery launched Operation Veritable and Operation Grenade, which were to clear out all German forces west of the Rhine in the north. Veritable was to kick off on February 8. The Canadian First Army, led by Lieutenant General Henry D. G. Crerar, would attack toward the southeast and the Rhine from Nijmegen, Holland. Leading the assault would be Lieutenant General Brian G. Horrocks's British XXX Corps and the Canadian II Corps commanded by Major General Guy G. Simonds. The attack would be a massive one, with more than 400,000 men involved.

For three weeks Horrocks had been wrestling with a major logistical task: bringing up the 202,367 men in his corps, together with thousands of guns, tanks, and vehicles, without being detected by the

The Roer River Dams

Germans. He had apparently succeeded, and now the woods in the vicinity of Nijmegen were crammed with soldiers and vehicles.

Nicknamed "Jorrocks," General Horrocks was thin and almost as-cetic-looking, features that belied his determination and war experiences. He had been a prisoner of the Germans, seriously wounded, and involved in heavy fighting since the North Africa desert days of 1941. General Crerar, his boss, knew that Horrocks could be counted on when the chips were down.

On February 10, two days after Veritable was scheduled to be launched, Simpson's Ninth Army would initiate Grenade by forcing a crossing of the Roer in the vicinity of Jülich, then drive toward the northeast to link up with Crerar's force advancing to the southeast from Nijmegen. Big Simp would send a powerful force of 303,000 men into the assault. To Simpson's south, 100,000 men of Major General J. Lawton Collins's VII Corps (part of Hodges's First Army) would attack the Roer to protect Ninth Army's flank. In all, Field Marshal Montgomery's 21st Army Group would have 800,000 troops involved in Veritable and Grenade—the equivalent of 27 divisions, 15 of them American.

General Hodges, getting heat from on high, made no effort to conceal his dissatisfaction with the snail-like pace of the 78th Infantry Division toward the Roer dams. His artillery commander, Brigadier General Charles E. Hart, had assembled 40 battalions of artillery (780 guns) to support the attack. "I don't understand it," Hodges exclaimed to Hart. "Why can't this enormous artillery firepower blast a road from our present positions straight to the dam?"[1]

At 11:45 A.M. on February 8, Hodges put in a call to Clarence Huebner at V Corps headquarters. The First Army commander minced no words over the lack of progress in seizing Schwammenauel Dam. Huebner responded that Parker's 78th Division had run out of steam. "Seizing that dam is too big a job for one division," he declared.

Hodges, a World War I infantry captain who had won the Distinguished Service Cross in the bloody Argonne forest, was unsympathetic. The entire advance to the Rhine in the sector of the U.S. Ninth and First armies depended upon Schwammenauel being seized.

"I want that dam in the morning," Hodges declared. "How you get it is up to you."

At midnight, 360 miles to the east of Berlin, Adolf Hitler was holding a meeting with a handful of confidants. Meeting at midnight was nothing unusual for Hitler; he was essentially a night person, sleeping through much of the day and engaging in exhaustive harangues to his military and Nazi party functionaries until nearly dawn. Then Hitler would retire for several hours of sleep. On rising he would take a hot bath, receive injections from Dr. Morell, and be ready to go again. At this hour his movements were slower, a trace of his sparkle had evaporated, but he still was alert and in control, after a fashion, of the German armed forces.

For five years Adolf Hitler had been saddled with an enormous burden that would have crushed most other men. He had assumed into his own person the functions of German Chancellor (head of state), Chairman of the National Socialist (Nazi) Party, Commander in Chief of the Wehrmacht (armed forces), and Commander in Chief of the Heer (army). It was a back-breaking load that customarily consumed a 16-hour workday.

Now the Führer was discussing the strategic picture, with Allied armies closing in from the west, south, and east. He admitted that "the situation is desperate." He was convinced that there was still a chance for salvation for the Third Reich, but only if the soil of the Fatherland was defended "inch by inch." "While we keep fighting, there is always hope," Hitler declared. "No game is lost until the final whistle."

He reminded his confidants that Germany was fighting a coalition of nations and emphasized that a coalition was not a stable entity and could split wide open "at any moment." "It exists only by the will of a handful of men," he exclaimed. "If the warmonger Churchill, who is conducting a holy war against the Third Reich, were suddenly to disappear, everything would change in an instant!"[2]

As Hitler warmed to his subject, he spoke of great warriors in Germany's past, of the goals and triumphs of the Nazi Party he had founded. He reminded listeners of the strategy and victories of his idol, Frederick the Great of Prussia, and how that fabled soldier had triumphed over coalitions when all seemed lost. In an excited voice he exclaimed, "We can still snatch victory in the final sprint! May we be granted the time to do so!"

Time. That was what Hitler needed. Time to turn out more of his *Wunderwaffen* (wonder weapons), which would reverse the entire course of the war. Time to produce hundreds of his revolutionary Messerschmitt jet fighter planes, which would render obsolete the Allies' propeller-driven air armadas and sweep them from the skies. A small number of the new-technology aircraft were already operational over the western front, and Hitler had rejoiced to a report that a single German jet had shot down 37 Anglo-American bombers.

At a facility code-named Dora, hidden in the rugged Harz Mountains of central Germany, 40,000 slave laborers under the direction of German technicians were toiling around the clock in underground factories to turn out the jets, which were far faster and more maneuverable than anything the Allies could put into the sky. Hitler would have taken enormous comfort had he known that his new jets, and the threat of hundreds more to come, were "scaring hell" out of the Allied air chiefs in England.

For the past six months, two of Hitler's Wunderwaffen—the V-1 robot bomb, a pilotless aircraft filled with explosives whose pulse-jet engine cut off over a target—and the much larger, liquid-fueled ballistic rocket known as the V-2—had been inflicting significant casualties on the civilian population of London. Now a brilliant German scientist, 34-year-old Dr. Wernher von Braun, was working on a multi-stage rocket that could put a satellite into orbit—or hit New York City, Washington, D.C., Philadelphia, and other cities on the American east coast. Hitler rubbed his hands in glee over the prospect. But his scientists needed time.

In Berlin, Major General Walther Dornberger, who possessed a rare combination of scientific brains and administrative skill, was in charge of developing these Wunderwaffen. It was the team of Dornberger and von Braun that had created the V-1 and V-2 weapons. Now Dornberger and his scientists were working on a remote-controlled missile that would destroy any Allied aircraft over Germany.

Also in the works was a new type of submarine that, it was hoped, would be able to travel underwater halfway around the world without surfacing, a revolutionary weapon that could play havoc with American and British battle fleets and merchant shipping. All of this would take time.

Because of these evolving weapons and his deep-rooted belief (backed by numerous historical precedents) that the Western alliance would yet fall apart, Adolf Hitler found grounds for optimism where other Germans saw none. But if there was to be a salvation for the Third Reich, his hard-pressed generals and their troops would have to buy time by fighting tenaciously for every inch of soil of the Fatherland.

Meanwhile panic reigned at SHAEF. A 2½-ton signals truck belonging to Major General Norman D. "Dutch" Cota's 28th Infantry Division had vanished from a street in Colmar, France, when left unguarded during the night. This was no ordinary GI truck. It held a top-secret encoding machine and related cryptography equipment. Had it been stolen by German agents?

In Allied-controlled Germany, France, Belgium, Luxembourg, and Holland military policemen halted every truck they saw. General Eisenhower demanded hourly reports on the mammoth search for the crucial equipment. Huge sighs of relief echoed through SHAEF when word was received that the truck had been located abandoned in a wooded area—with the encoding machine and the cryptography gear intact. Presumably the truck had been stolen by French civilians. Or had it? Two of three safes containing top-secret cryptographic information were recovered, but the third safe was missing.

3

An Angry Patton Threatens to Resign

General Brian Horrocks awoke an hour before dawn on February 8 at his headquarters at Tilburg, Holland. It was miserable weather, chilly and starting to rain. If there was anything the British XXX Corps commander did not want as the whistle was about to blow for the kickoff of Operation Veritable, it was rain. Rain would limit or curtail powerful fighter-bomber support, and bog down his masses of Cromwell and Sherman tanks and his "kangaroos" (armored troop carriers) in the muck. The rain that pelted the battleground at Nijmegen was a bad omen.

Horrocks pulled on a raincoat, went outside, and mounted a high platform that gave him a panoramic view of the Dutch plain. Minutes later there was an enormous roar behind him and jagged, lightning-like flashes illuminated the dark sky. More than 1,400 big guns, massed almost hub to hub, had begun to pound German positions. Far to his front General Horrocks could see countless orange flashes and black puffs of smoke as thousands of shells screamed in on the enemy.

Suddenly the deafening bombardment ceased. Hundreds of Horrocks's tanks and kangaroos lunged forward, slithering through the freezing mud.

At 9:25 A.M. the big guns opened up once more and laid down a "creeping barrage," advancing their impact area 100 yards every few minutes as assault infantrymen on foot moved forward. Then WP (white phosphorous) shells began to fall, and soon a thick curtain of smoke stretching across the front cloaked the Tommies and blinded the German observers. Battalions of four different British divisions began edging toward the ominous, thickly wooded Reichswald, meeting only sporadic resistance. General Horrocks was elated. It appeared that Veritable was off to an auspicious beginning. Abruptly, though, his euphoria switched to concern, then to alarm. Horrocks's tanks were bogging down in the mud.

Within hours, Veritable was in chaos. The steady rain had continued to pour down in sheets, turning fields into quagmires. Enormous traffic jams developed on the roads. Tanks, trucks, jeeps, and guns were stalled bumper to bumper for miles on the main supply road leading from Nijmegen to Cleve. The attack ground to a halt.

At his headquarters west of the Roer River, General Simpson was enduring the agonies of the damned. Reports had been flowing in from Horrocks's front—all of them dismal. Word from the south was that the ponderous Schwammenauel Dam was still in German hands. Simpson's assault over the Roer to link up with Horrocks's troops—Operation Grenade—was to jump off at dawn the next day, but if he sent his assault troops over the river and the Germans opened the floodgates at Schwammenauel, the Roer Valley would be inundated for widths of up to a mile. Then the Germans could move in against Simpson's marooned assault troops and cut them to pieces.

The success of the British attack to the north depended on Grenade being launched by Big Simp's men. But could he take the chance and see his stranded spearheads wiped out on the far side of the Roer? On the other hand, if he called off Grenade on the morrow and the Germans did not flood the valley, he would be summarily sacked, sent home in disgrace, and forever branded as a timid battle commander who had jeopardized the entire massive operation to clear the Germans from the west bank of the Rhine. All but one of five American

27

corps commanders to be involved in Grenade urged Simpson to cancel.

Simpson retired to his office—alone. For long minutes he agonized over the biggest decision of his career. At 4:00 P.M., 14 hours before he was to assault the Roer, Simpson flashed the word: "Postpone Grenade!"

At his headquarters at Bonn, on the Rhine, Field Marshal von Rundstedt had sprung into action on receiving reports of Horrocks's massive offensive. The German commander in the west began stripping troops from in front of Simpson's stalled Ninth Army along the Roer and rushing them to the north to confront the British and Canadians. One of the most grueling, bloody battles of the war erupted in the muck and mire of eastern Holland. Altogether, von Rundstedt sent nine divisions northward, while General Simpson was forced to sit by in idleness, monitoring proceedings, unable to swing his "right hook" to link up with Horrocks because of the very real threat posed by the Schwammenauel Dam and its 81,000 acre-feet of pent-up water.

At virtually the same time that German commanders were reading von Rundstedt's signal ordering a northward shift of nine divisions from in front of Simpson and Collins's VII Corps, American and British leaders were eagerly perusing the same top-secret message. In some cases, the Allied commanders read von Rundstedt's signal before it reached the Wehrmacht addressees. Only a handful of top generals and staff officers among the Western Allies were aware of the "spy" in the enemy's higher echelons of command. That "spy" was code-named Ultra, an ingenious British program that intercepted and decoded top-secret German electronic signals. Since the beginning of the war Ultra had furnished British and American military leaders with a continual flow of details on the German command structure, and on the strength, location, and morale of enemy units, often down to battalion or even company levels.

Ultra was one of the most closely guarded Allied secrets of the war. It had its origin in 1939, just after hostilities erupted in Europe. British intelligence, with the aid of Polish agents, had stolen a precise copy of the highly secret and complex German coding machine named Enigma. So sophisticated was the Enigma machine that Adolf

Hitler and the Wehrmacht considered its code unbreakable. A team of Great Britain's leading cryptanalysts and mathematicians, working for months under the most intensely secret conditions, solved the "unbreakable" Enigma code with the aid of another highly complex machine. From that point on, top-secret German signals were intercepted by the British.

The nerve center of Ultra was a large, stone Victorian mansion located 40 miles north of London, just outside the town of Bletchley Park. At this highly secret center, called Station X, German coded messages were intercepted, sometimes evaluated, then encoded again and transmitted to Allied commanders on the western front. Only a few officers at SHAEF, Omar Bradley, Bernard Montgomery, army commanders, and top intelligence officers were privy to Ultra signals. At corps command and below, the existence of Ultra was not even known.

At each of the American or British headquarters authorized to receive Ultra's pirated signals was a wireless truck manned by Royal Air Force and Royal Corps of Signals officers, whose mysterious function was a source of unending speculation among those not in the know. These trucks and personnel were known as SLUs (Special Liaison Units). When the officer in charge of each SLU received a message from Station X at Bletchley Park, he would slip out of the truck with the decoded signal and personally hand it to the authorized American or British general or intelligence officer. Once it was read, the message was returned to the SLU officer, who would return with it to his truck and promptly burn it. No written record of the German top-secret signals was allowed in the field.

Throughout the war, Adolf Hitler never lost faith in his "unbreakable" Enigma code. On occasion, when speculation arose in the Wehrmacht high command that the Western Allies were learning in advance of German plans, Hitler accused "spies and traitors" in the German general staff with leaking the information to the West.

It was nearing 6:00 P.M. on February 9 when Colonel John G. Ondrick, commander of the 309th Regiment of Parker's 78th Infantry Division, gave a curt order to his 1st Battalion leader, Lieutenant Colonel Robert H. Schellman: "Seize Schwammenauel Dam . . .

tonight!" It would be an exceedingly difficult operation. Schellman's "dogfaces," as the American foot soldiers proudly called themselves, would have to fight through thick, pitch-black forests and deep, treacherous ravines, against Germans ordered to hold at all costs.

The 1st Battalion jumped off immediately. Lieutenant Colonel Schellman soon lost contact with his companies, as radios refused to function in the ravines. Heavy automatic weapons fire raked the advancing infantrymen, but they pushed forward. The battalion split, part heading for the high ground at the crest of Schwammenauel Dam and the remainder descending to the base of the enormous structure. At 9:15 P.M. one company edged out of a patch of woods and found itself on a bluff peering down through the darkness at the shadowy outlines of the huge dam.

Now the Germans were aware that the Americans had reached Schwammenauel. The race was on: Could Schellman's men seize the dam before the enemy could blow it up?

At 11:00 P.M. a five-man team of the 303rd Engineer Battalion was rushed forward to launch a search for explosives. A layout of the massive dam had been studied in detail the previous day. Crouching low, hearts pounding, the engineer group, accompanied by a few riflemen, began dashing across the top of the dam, a structure as high as a 25-story building and as long as three football fields. Bullets hissed past the running Americans.

Suddenly the engineers lurched to a halt. A few more steps and they would have plunged to their deaths, as the Germans had already blown a portion of the spillway in front of them. Flopping face downward, the nervous engineers and riflemen hurriedly discussed a new course of action. Streams of tracer bullets, fired by Germans at the far end of the dam, continued to split the blackness around them, adding to their sense of urgency. The plan had been for the little band to charge completely across Schwammenauel, then descend to a tunnel running through the structure where explosives would most likely be placed. Now there was only a single alternative—a perilous slide down the face of the dam to the tunnel 250 feet below, a feat requiring the skill of a mountain climber and most likely the endurance of an Olympic decathlon athlete. If there shadowy forms were spotted in the descent, they would be picked off the face of the dam by enemy marksmen.

Getting to their feet, members of the search team went over the guardrail and started down the face of the structure. They negotiated the arduous descent undetected, and at the bottom landed on and captured six startled German machine gunners who had not expected to be pounced on from above. The engineers hurried into the tunnel and began frantically searching for explosives.

Inside the dam the atmosphere was stifling. Soon the Americans were sweating profusely and breathing with difficulty. While the battle for control of the dam continued to rage outside, the engineers groped through the tunnel, all too aware that 22 billion gallons of water were straining at the dam and could burst over them at any moment. Even as they pressed their search, an already lighted fuse might be burning toward an enormous charge of dynamite somewhere in the bowels of the massive dam. The minutes ticked past . . .

Much to the astonishment—and relief—of the search team, the tunnel was devoid of prepared charges. The engineers emerged into the night, the frigid air outside a welcomed change from the suffocating pressures inside the tunnel. Now a thorough combing of the outside of the dam was begun.

Two hours later the inspection was completed. The Americans swarming over mighty Schwammenauel in the blackness breathed more freely; the booby traps and delayed charges expected to be in abundance were not present. At 2:00 A.M., eight hours after jumping off, Lieutenant Colonel Schellman radioed the 309th Regiment commander: "Schwammenauel secured!" But the absence of explosives had not been a monumental blunder by the customarily thorough and methodical Wehrmacht. The Germans had done precisely what they had intended: a 13-foot-wide hole had been blasted in Schwammenauel's tough hide, permitting a steady stream of water to gush into the Roer Valley in front of the poised U.S. First and Ninth armies, forcing Grenade to be postponed for at least two weeks.

Ninety miles to the north that day, General Brian Horrocks's men tried to push forward on the second day of Veritable but were thwarted by knee-deep mud, thick minefields, and tenacious German resistance. The British XXX Corps attack had turned into a naval operation of sorts. The main supply road from Nijmegen to Cleve was underwater, and four ferries were put into operation to bring ammunition and other vital needs to the men on the front lines.

That same day, General George Patton, grumpy for days over the secondary role assigned to his Third Army, was wearing what his devoted staff called his Number-3 frown. A call came from his boss, Omar Bradley. In his high-pitched voice, the 12th Army Group commander inquired, "Georgie, how soon can you go over to the defensive?"

On several occasions George Patton had told Bradley, "I wish to goddamned hell all I had to fight was the Hun instead of battling Ike and Monty half the time!"[1] Now Patton's face rearranged itself into a Number-4 frown and flushed with anger. With Grenade stymied by the Roer flood and Veritable bogged down, Patton knew that Third Army was the only Allied force advancing on the entire western front, but the SHAEF strategic plan for vaulting the Rhine called for only Field Marshal Montgomery to be attacking. Patton exploded. Turning the air blue, he shouted over the telephone that he was the oldest combat commander and the most experienced in the American army, and swore that he would promptly resign if forced by SHAEF to go on the defensive.

Minutes later, Bradley called back, apparently in an effort to smooth the feathers of his irate army commander. "Monty's bogged down in the north," Bradley exclaimed with only a trace of an I-told-them-so tone. He speculated that SHAEF would now slip the ball to Patton and let him carry it to the Rhine and beyond. Patton snorted that he doubted such would occur. The two generals agreed on one point in the conversation: Veritable and Grenade had been major Eisenhower blunders; the supreme command had been unduly influenced by Field Marshal Montgomery in developing the master plan.

Shortly after Patton had been given the "aggressive defense" role for Third Army on February 1, he had wangled authorization to "continue the probing attacks now in progress." Secretly, he was hoping to develop a "probing attack" into a full-fledged breakthrough that would plunge Third Army all the way to the Rhine, 75 miles away, head of Field Marshal Montgomery, who was bogged down in the mud to the north. But the road to Hitler's historic moat was proving to be a bloody one, as Third Army had to attack over some of the most inhospitable terrain in Europe—forested, hilly, crisscrossed with deep ravines and gorges—a region known as the Eifel.

Patton was deeply disturbed by the heavy casualties being inflicted on his troops. But German Army Group B facing him in the Eifel, although resisting ferociously, was suffering in equal measure. Under the relentless strain of trying to ward off Patton's blows with decimated units and an absence of airpower, raucous disputes erupted among high-level German commanders.

Lacking the strength to counterattack Patton's deep penetrations at Prüm and at Bitburg, General of Infantry Hans Felber, the battle-tested leader of XIII Corps, contacted his immediate superior, General of Panzer Troops Eric Brandenberger of Seventh Army, with a request to withdraw to more defensible positions. Brandenberger's reply was a succinct *"Nein!"* even though he personally favored such a pullback. He himself had recommended withdrawal the previous day to Field Marshal Walther Model, leader of Army Group B and a Nazi zealot recognized by the Allies as one of the Wehrmacht's most capable tacticians.

His monocle in place and his face stern, Model had heatedly refused Brandenberger's request to withdraw. Hitler's standing order was "Don't give up a foot of ground!" But the diminutive Model would have angrily rejected Brandenberger's request anyhow. He had been outspoken in condemnation of the Seventh Army commander for his failure to repulse Patton's drives on Prüm and Bitberg, and was in a mood to fire Brandenberger.

Walther Model, who at 54 was Germany's youngest field marshal, was regarded by many old-line Prussian generals, such as Gerd von Rundstedt, as somewhat below their lofty aristocratic status—Model's father had been a schoolteacher, a commoner. Model had gained his exalted rank partly through his achievements on the battlefield but also through his fanatic devotion to his Führer. When Hitler had miraculously escaped being killed in the July 20, 1944, bomb plot, Model had been the first officer to send a message to the shaken leader pledging his undying loyalty. Come what may, Walther Model could be counted on to fight to the end.[2]

That day in the Reich Chancellery in Berlin, Albert Speer, the 37-year-old production genius from Heidelberg, was breaking alarming news. Hitler had great faith in Speer, the architect-turned-industrialist, who for several years had been the guiding spirit behind the Third Reich's armament manufacturing.

"*Mein Führer*, once the Anglo-Americans take over the west bank of the Rhine opposite the Ruhr, the disorganization of our main coal and steel region and the closure of our river transport lines will mean an economic collapse in from four to six weeks," Speer told a stone-faced Hitler.[3]

It was hardly the kind of news Hitler had been hoping to hear from the mastermind of his war production. But Speer's revelation convinced him that his embattled armies would have to slug it out with the Allies as far as possible from the Rhine River and its adjacent Ruhr industrial basin.

Late in the afternoon of February 8, sultry-voiced Arnhem Mary was cooing over Radio Berlin to British and Canadian soldiers fighting in the mire and floodings in the north. Arnhem Mary was an old friend of the Tommies, and they had given her the name several months previously during the heavy fighting in Holland following a massive Allied airborne drop and landing. Presumably to establish her credibility, Arnhem Mary ticked off the precise Canadian and British units in action and their positions along the battle line. Then she proceeded with her daily task of trying to foment dissension among the Allied fighting men.

"The Yanks are fighting hard down south," she cooed, "and the British are attacking through the Reichswald. What are the Canadians doing? They're eating hamburgers in Nijmegen."

Indeed the British were battling fiercely in the thickly forested Reichswald and, despite the miserable terrain and weather and a tenacious German resistance, were edging forward. The arduous progress had buoyed spirits of some at headquarters of Horrocks's XXX Corps to the point that the commanding general that day laid a bet with an aide that the British would seize intact a bridge over the Rhine. Few others in corps headquarters, however, were that optimistic.

4

"Sleeper Patrols" Among the Enemy

Tension was mounting along both sides of the turbulent but receding Roer River. On the American side of the stream, a thunderbolt was waiting to strike. In and around nearly every building, barn, ditch, and highway underpass there were men and vehicles: tanks, tank destroyers, artillery, mortars, signal facilities, military police, engineers, riflemen. Along the banks of the Roer, American dogfaces moved about silently, under a cloak of darkness where possible. They made their way in shifts to and from outposts almost at the water's edge. At the outposts, the stiff and miserable machine-gunner or rifleman remained motionless for long hours. He dared not move. Only a few yards away, across the cold waters of the Roer, equally miserable German soldiers were hunched in icy foxholes, ready to fire machine-gun bursts into any moving object on the American side.

Most of the time it was quiet, eerie, ghostlike. Here and there the sharp crack of a rifle pierced the frigid air. "Incoming mail"—enemy artillery and mortars—occasionally screamed into the American side. More often it was "outgoing mail." In the forbidding gray overcast the putt-putt-putt of a German robot bomb could be heard, heading for the Allied supply port of Antwerp or the center of London.

"Fritz [the stereotyped German soldier] over there gets his jollies on hearing those goddamned flying bombs go over," an American machine-gunner in the 84th Infantry Division remarked to a comrade. "But they won't do Hitler any good in stopping this steamroller." His buddy nodded his head in silent agreement.

Across the Roer from Major General Terry de la Mesa Allen's 104th (Timberwolf) Infantry Division lay the bleak shambles of Düren. After months of heavy bombings and artillery pounding, the once stylish town was a twisted mass of stone, concrete, and timber, pockmarked and gutted. On occasion a German would steal molelike from his underground refuge among the rubble piles, only to be greeted by an artillery or mortar barrage directed by American observers in church belfries.

Life on the Roer during the lengthy wait for the Roer dams to be captured and then to be drained was a mixture of boredom, periodic danger, comedic episodes, deadly serious patrolling, and a feeling of "Let's get it to hell over with!" Most of the GI fighting men along the Roer lived like moles, buried deep as protection against German artillery and mortars and the occasional night bombings by lone Luftwaffe aircraft known to American fighting men as Washing Machine Charlie. Some lived in comfort.

Men of the 414th Regiment in the sector north of Düren found that the Germans had wrecked the power lines and had taken all candles and oil lamps with them, a deed classified by one irate dogface as "a dastardly war crime." A water-driven generator was found in a ruined foundry, and Privates First Class M. F. Redecker and George Martin took advantage of the lull to rewire the little village and put the generator into operation. As a result the GI "tourists" in Lendersdorf were supplied with electric lights and radios, wholly unanticipated frontline luxuries.

Staff Sergeant Arthur E. Williams of the Timberwolf Division periodically updated a curious little "black book" he had carried for weeks. Comrades never needled him as to how many blondes and how many brunettes were listed in it. This little black book had been scrupulously maintained by Sergeant Williams since his first night in combat, when his closest friend had been killed. Since then he had sworn eternal vengeance against the Wehrmacht. In his book he listed "weapon used," "name of victim" (if he had time to search the German body), and "circumstances" surrounding the kill. He had listed 23 entries and told comrades there were "plenty more pages available."

The flood-caused stalemate along the Roer resulted in many strange minor duels. Bored observers in a church steeple in gutted Merken, opposite Düren, spotted several German grenadiers across the swollen river and opened fire on them with 81-millimeter mortars. The Germans heard the popping sounds of the mortars going off and dove for ditches and foxholes until the shells exploded around them. Unscathed, the *Feldgrau* ("field gray," meaning the German regular soldier) climbed out of their cover and continued with their work. The American mortarmen fired again, with the same results. This process continued for more than two hours.

Staff Sergeant Robert Barton and his mortar squad of the 413th Infantry Regiment finally timed a shell in flight. It took 22 seconds for the shell to leave the muzzle and explode. Barton fired a round, waited 22 seconds, then let fly another. The Germans ducked the first round, having heard its popping noise. But the sound of its blast muffled the report of Barton's next rounds, so the Feldgrau in the cat-and-mouse game climbed out of their holes once more. In 22 seconds the rounds they had not heard arrived and burst among the Germans. The "game" had ended.

Cold and bone-weary men of I Company of the 413th Infantry had moved into a gutted German paper plant. Waist-high piles of debris covered the floor of the roofless structure. "Make yourself at home!" a captain called out dryly. The men, knowing this would in fact be their home for a period of time, set out to do just that. Private Delbert Erwin masterminded a project that turned concrete storage bins into warm, dry, little sleeping cabins. "This is a model for postwar, low-cost housing," Erwin remarked after surveying his remodeling job. Private First Class Andrew Posey was the "foreman" on the renovation project, supervising the building of complete new bathrooms, with tubs, heaters, and running water. A 60-millimeter mortar squad built a clubhouse, which they cristened Club 60, complete with writing table, pinups of curvaceous Hollywood stars, and even a working telephone. "The only trouble with the goddamned telephone," complained one straight-faced mortarman, "is that all we can call are some other goddamned filthy dogfaces!"

The OP (observation post) of Easy Company of the 414th Infantry Regiment was in a house at the edge of the Roer within stone-tossing distance of Germans burrowed in underground bunkers on the far side. In the quiet of night the GIs could hear enemy soldiers gossiping. The living room was furnished in perfect taste, and GIs spent long hours there reading while seated in overstuffed chairs. Pinup pictures of Rita Hayworth, Betty Grable, and Hedy Lamarr adorned the walls. Hours were devoted to heated disputes as to the various physical attributes of one star as compared to another. Only the broken windows furnished mute evidence that a war was being fought there. A stove kept the living room cozy and comfortable. "It's a pretty good set-up here," one rifleman observed to a visiting reporter. "The only reason some guys don't prefer it to a rear area rest center is on account of the patrols to the other side of the Roer."

Patrols to probe the enemy on the far side of the stream were an almost nightly occurrence along the Roer. The nocturnal jaunts into enemy-held terrain were full of peril. Men trying to cross the river in the darkness were swept off their feet by the current, some dragged under to their deaths. A three-man team attempted to ford the Roer at night and drive stakes for guide ropes to help assault boats gain the far bank when Grenade kicked off. An engineering officer and one man were never heard from again.

On a blustery raw night, Platoon Sergeant Karl H. Stelljes led a patrol into the murky, swift-flowing waters of the Roer with orders to seize prisoners for interrogation. Wading as noiselessly as possible, the 104th Division patrol reached the far bank, paused to listen, then stole onward. Sergeant Cecil E. Ross set up a BAR (Browning automatic rifle) along the bank as the others edged forward in a skirmish line. Stelljes and his men had advanced only 20 yards when they potted a trench. Fresh muddy tracks could be discerned, which indicated the Germans had been using the excavation recently.

Sergeant Donald R. Weishaupt and three men concealed themselves in a thicket to cover the trench while Stelljes and two others moved forward, slipped down into the trench, and began cautiously walking along it, weapons at the ready. The only sounds were the faint squishing noise of their boots in the oozy mud and the howling of the wind. Suddenly, the stillness was shattered. Out in the darkness a German burp gun (machine pistol) began sending streams of tracers over the Americans in the trench.

Stelljes began slipping back down the excavation to get Sergeant Weishaupt and his other men. Private First Class Emil Zegerheim fired a green parachute flare into the black sky, a signal to the Americans on the other side of the Roer that mortar fire might be needed.

Returning with Weishaupt and his three comrades, Sergeant Stelljes and the rest of the patrol once more moved forward along the bottom of the trench. After going about 40 yards more, the night air was split by a sharp challenge: "Halt!" The Americans froze. Stelljes pulled out a grenade and pitched it in the direction of the German voice. Another patrol member tossed a white phosphorous grenade. Stelljes, with Weishaupt at his heels, then charged forward and fired Tommy gun bursts blindly into the enemy position.

A German voice called out, "*Kamerad! Kamerad!* [I surrender!]" The two Americans could detect the silhouettes of two enemy soldiers emerging from a dugout at the end of the trench. Quickly, one German swung his rifle in the direction of Weishaupt. Sergeant Stelljes did not have time to raise his Tommy gun and shoot. He hurled the weapon at the German, then leaped on him, wrested his rifle away, and brought the stock crashing down on the enemy soldier's head.

"Okay, let's get the two Krauts and get the hell on back," Stelljes called out in a stage whisper. The patrol recrossed the Roer with the prisoners, who were promptly hustled away to be interrogated at 104th Division headquarters.

It was nearing midnight on February 13 when Sergeant William M. McIlwain and three other members of his Timberwolf patrol were creeping and slipping through the mud on the German side of the swollen Roer. McIlwain and the others had volunteered for this "sleeper patrol," but now he was having second thoughts. "I must be crazy," flashed through his mind. McIlwain and his comrades would hole up during the day and observe German positions and guns, send the information back by pigeon, then slip back over the river the following night.

McIlwain was thoroughly miserable—soaked to the skin, shivering from the cold, exhausted from the physical and mental strain. Suddenly, Platoon Sergeant Flores, the patrol leader, put up his hand. The column halted in place. McIlwain felt his heart thumping rapidly. Flores waved for his men to join him, pointed downward, and by the muted rays of a moon darting in and out of clouds McIlwain and the others saw a German soldier sleeping soundly in a muddy foxhole. The German's snores seemed to carry far across the landscape, making the Americans nervous.

Not a word was spoken. The patrol members knew their business. Private First Class McCain whipped out his razor-sharp trench knife, knelt beside the enemy soldier, and held the blade to the throat of the unsuspecting foe. Whether the German lived or died depended upon whether he happened to awaken. The other three in the patrol crept softly away, soon followed by McCain. The slumbering Feldgrau was left unharmed.

McIlwain and the others trudged on through the darkness for the next half hour, deeper into German territory. Again the column was brought to an abrupt halt. Sergeant Flores had spotted antipersonnel mines. The patrol gave the field a wide berth, and from that point on the Americans lived in dread of the thought that each step might activate a mine and blow off a foot or leg. It was a terrifying reflection: if anything went wrong, there was nothing comrades could do to help, and the victim would suffer excruciating agony until he bled to death.

Cautiously moving along in single file, McIlwain and the others neared a house that had been picked out in aerial photos as the patrol's observation post. The column froze. A shaft of light was seeping out of one window, and from inside came loud German voices raised in a avorite ballad of the Feldgrau, "Lili Marleen." Obviously, a raucous ¤arty was going full blast.

"I can't attend the party," a patrol member whispered. "I left my ¤uxedo back across the Roer."

The scouting force pulled back some hundred yards and climbed nto shell holes to watch and listen. The holes were partially filled ¤with water. For an indeterminate period the Americans lay freezing in ¤he icy wetness, blaming themselves for their miserable, perilous ¤light. Their light field jackets were soaked. "To think I left a warm, ¤ry cellar to volunteer for a place like this!" Sergeant McIlwain re-¤lected. "If I get back I'm going to apply for a Section Eight," (the ¤army designation for a psychological discharge).

Eventually dawn arrived, but instead of a warming sun, the shiver-¤ing Americans were greeted by a murky sky, heavy winds, and rain. One of the patrol members reached into his gas-mask cover and pulled ¤out one of the two pigeons he had been carrying. A terse message was ¤attached to its leg and the bird was released. It headed deeper into German territory, then got its bearings, banked, and flew off toward ¤ts roost on the American side of the Roer. McIlwain and the others ¤watched breathlessly until the pigeon was out of sight.

With daylight, the four Americans peered over the rims of the shell ¤holes and saw German soldiers starting to move about on all sides. Suddenly, a new peril surfaced: "friendly" shells from across the Roer ¤began exploding around the patrol's holes, exposing them to the fire ¤that was to have protected them in the designated house.

Throughout the day, the Americans observed German activities ¤and spotted several artillery and machine-gun positions. Despite the ¤tenseness of the situation, McIlwain realized that he was enjoying ¤spying on the unsuspecting Germans from right in the midst of their ¤positions. But he and the others greeted the arrival of dusk with relief. It had been the longest day of their lives.

It had also been a productive 24 hours. The patrol had counted ¤Germans, spotted a 120-millimeter gun, sketched trench networks, ¤and marked the large strip of shoreline saturated with mines into ¤which they had nearly stumbled. This information was sent back over

the Roer by the second carrier pigeon—a communications techniqu●
reminiscent of wars of yore. But there was good reason for the feath●
ered messenger: if the patrol didn't make it back, at least its intel●
ligence findings would get there.

Stalking cautiously through the night on their return trip, Sergean●
McIlwain and his comrades had nearly reached the Roer when aler●
enemy machine-gunners opened fire. With bullets hissing past them●
the Americans dashed for the river and crossed to the west bank.

Three nights before the Roer was to be assaulted, intelligence officers
in Allen's 104th Infantry Division had grown alarmed—a not un-
familiar state of mind among those engaged in trying to divine the
enemy's dispositions and intentions. For weeks the Timberwolves had
sent numerous patrols over to the German side, and a composite picture had been constructed of German wire barriers, minefields, machine-gun nests, and trench networks. Now, on the eve of Grenade,
compelling evidence indicated that the enemy opposite Düren had reorganized his defenses and might be bringing in a complete new division.

That night a German soldier waded the Roer and surrendered to a
Timberwolf outpost. He was rushed to division headquarters, where
he was intensively interrogated by Captain William Stelling, who
hoped to obtain additional information about the German positions on
the far side of the Roer. Time was running out. The attack would be
launched soon. The prisoner tried to be cooperative, but pleaded that
he had dashed through the river lines so quickly that he did not have
time to talk with any Germans on the far bank.

Desperate for current information and convinced that the German
was sincere in his claims, Captain Stelling was struck with an inno-
vative plan: he would send the prisoner back over the swollen Roer
that night and have him walk up and down the enemy-held shore and
ask each German he encountered the designation of his unit. The
POW agreed to carry out the task.

But Stelling had second thoughts. Was this German telling the
truth? Had he actually risked a hazardous crossing to surrender, or

was he on a fact-finding mission for the Wehrmacht? Maybe it was all part of an elaborate *ruse de guerre*.

"If you're lying, you're a goddamned dead Kraut!" the German was told, somewhat illogically, as he set out on his mission shortly after midnight. Lieutenant Burleigh Sheppard and six of his engineers took the prisoner to the water's edge and ferried the German back to the enemy side of the river, then returned to the west bank. The plan was for the POW to obtain the information, then swim back to the American side.

Captain Stelling waited anxiously for word that his converted spy had returned. He kept glancing at his watch . . . 1:00 A.M. . . . 1:30 . . . 2:00 . . . 2:30 . . . still no sign of the German. Stelling's field telephone suddenly jangled impatiently, and as he picked up the instrument and listened to the voice on the other end a broad smile spread across his face. It was an outpost along the Roer bank. The spy had just climbed out of the river after swimming back across.

Stelling soon learned that the information the German had obtained was far in excess of what he had hoped for. The enemy soldier had checked in with several German outposts along the Roer and confirmed that a new division indeed had recently moved into position. In addition, the division's designation and new position were obtained. The information would prove to be precisely accurate.

Bizzare events surfaced on occasion in the eerie silence of the stalemate along the Roer. A number of American artillery observers, presumably in control of their faculties, reported to fire-control centers in the rear that they had heard the sound of train whistles in Düren. A train in the gutted, deserted town seemed impossible. But when the reports continued periodically, the artillerymen attached a name to the phenomenon—the Düren Express. Several times batteries fired on supposed train locations, although nearly every square yard of the town was pockmarked with bomb or shell holes that would have rendered the likelihood of an intact railroad track virtually nil.

Several hundred feet above the Roer on a gray, overcast day, Lieutenants Everett E. Jones, observer, and Robert C. Dwelly, pilot, completed a routine air reconnaissance mission for the 385th Field Artillery Battalion and headed their Piper Cub for home. Two miles from its base, the Cub was attacked by six ME-109 fighter planes. Lieutenant Dwelly put the light aircraft into a steep dive as three of the Luftwaffe planes opened up with machine-gun fire.

Dwelly leveled out at treetop height and flew up a shallow valley. The first German fighter that had dived on the Cub was unable to pull up and crashed into the ground, exploding in a huge burst of orange flame and black smoke. Dwelly and Jones were officially credited with the destruction of a Luftwaffe pursuit plane and the next day stenciled a swastika on the unarmed Cub's fuselage.

At his palatial SHAEF headquarters in Versailles, on the outskirts of Paris, Supreme Commander Dwight Eisenhower was meeting with the press on February 20. Just as tension and anxiety were mounting by the hour among assault troops along the Roer, it became obvious that the Big Showdown was about to erupt; Eisenhower's somber demeanor reflected a similar mood at the highest level of command. Without his customary preliminary pleasantries, he told reporters, "One of the decisive battles of the war is about to be fought on the Cologne Plain."

Pointing out that the purpose of Grenade was to destroy all German power west of the Rhine, Eisenhower said that the Allies would storm over the Rhine and push on to a juncture with the Russians at some undisclosed locale. "There is going to be no cessation in aggressive action on the [western] front," he exclaimed. "We are going to fight as violently as we are able from now on."[1]

Behind the swollen Roer River, General Bill Simpson had been gnawing his fingernails in frustration for two weeks. D-day for Grenade had been postponed day after day. By February 20 the flooded Roer was receding, and engineers advised the Ninth Army commander that the big Schwammenauel Dam would be emptied by February 24. That was the news Big Simp had been anxiously awaiting. He set D-day for February 23, one day before the dam ran dry, in the hope of gaining a degree of surprise.

Simpson's powerful force would assault the Roer along a 15-mile

retch, from just below Linnich on the south to four miles above
ülich on the north. Spearheading Grenade from Ninth Army would
e assault battalions of Major General Charles H. Gerhardt's 29th
nfantry Division, veterans of Omaha Beach in Normandy; the 30th
nfantry Division, led by Major General Leland S. Hobbs; the 102nd
nfantry Division, under Major General Frank A. Keating; and the
4th Infantry Division, commanded by Major General Alexander R.
olling. In addition, Simpson would have the heaviest concentration
f guns ever assembled by an American field army—2,038 artillery
ieces, supplemented by the guns of countless tanks and tank destroy-
rs, plus hundreds of mortars and antiaircraft weapons.

Jumping off at the same time on Ninth Army's right flank would be
General "Lightning Joe" Collins's veteran VII Corps. Collins would
ead with his 8th and 104th infantry divisions, with the 3rd
Spearhead) Armored Division in reserve.

Ultra had revealed that von Rundstedt had virtually denuded the
Roer River sector to send troops northward to halt Montgomery's
Veritable. Only some 40,000 men and 75 artillery pieces would be on
and to oppose the awesome American assault force. Yet Simpson, a
6-year-old former polo player from Weatherford, Texas, who was
nown in his younger days as "Texas Bill," was not complacent about
he final results. An assault river crossing was one of the most perilous
nd difficult operations, and Simpson knew that the Germans would
e deeply dug in, having had months in which to zero in mortars and
rtillery. The Roer could well run red with blood—American blood.

t was nearing midnight on February 22. H-hour was less than four
hours away. Americans along a 25-mile stretch of the still-angry Roer
River were plagued with a curious mixture of foreboding over what
ay ahead and a vague sense of relief that the interminable waiting was
nearly over. The moon was especially bright—too bright for the infantrymen and engineers who would be the first to cross the Roer.

"Hell, the goddamned Krauts can see us coming a mile away!" a
nervous infantryman across from Düren howled.

There wasn't much for the assault troops to do now except huddle
in basements and other shelter and await H-hour. All plans had been

made and troops were in position near hidden assault boats in which they would try to cross the Roer.

In the cellar of a battered house near the river, the command group of a 104th Infantry Division battalion were sprawled about in near silence. Captain Robert Neilsen looked at his watch. His face was etched with the wrinkles of fatique and his eyes were red from lack of sleep and tension. "It will be a lifetime between now and three o'clock," he commented softly to a comrade. There was an undercurrent of tenseness in the little room. A truck idled by in the moonlit darkness outside, a tiny noise but one that caused each man to suddenly look up.

Major Ray Waters broke out with a forced laugh. "When I walked down the road a few minutes ago I tried to tiptoe. Those noises sound awfully loud at night with the Krauts just across the river." No one commented.

A field telephone in the cellar jangled insistently. Captain Neilsen picked up the receiver, listened briefly, and said, "Okay, I'll try to check it out." Then the young captain from Bloomington, California, looked up at the grim, searching faces of the others. "An outpost by the water says some unidentified figures are slipping along the river bank on *our* side." Neilsen began checking with nearby units on the telephone.

"We'd feel silly as hell if the Krauts came over and stole our boats," Major Waters said.

Finally Neilsen hung up the receiver and called out, "It's okay. Some of our guys are setting up smudgepots along the bank in case we need a smoke screen."

"What time is it now?" a nervous voice called from across the cellar.

"Four minutes later than when you asked last time," replied another.[2]

In an adjoining room, Lieutenant Colonel Fred Needham, a battalion commander from Auburn, California, was sleeping soundly, undisturbed by the nearby chatter. He was resting up for the hard grind just ahead.

Near the banks of the Roer in another cellar, Captain Jerry Hooker of St. Petersburg, Florida, was idling away the time. He would lead one of the assault companies in the crossing. Hooker looked far too young to be leading a company of infantry—even younger than his 24 years. All around him headquarters aides were slumped on the floor, trying to look unconcerned. There was little talk. What there was was brief and to the point. It was 1:30 A.M. H-hour was two hours away.

Captain Hooker pulled out a bottle of Scotch with a couple of swallows remaining in it. Booze had been hard to come by along the embattled Roer. "I've been saving this," the company commander observed. He took a drink and handed the bottle to Lieutenant Francis Ahrnsbrak of Marshall, Oklahoma, seated nearby, saying. "Here, have one, Francis."

"Thanks," replied the lieutenant, draining the rest. "I wish I had those nine bottles I ordered."[3]

A shell crashed nearby and the little house shuddered.

5

Assault Over the Roer

An eerie quietude hovered over the moon-bathed Roer River at 2:44 A.M. Suddenly the hush was shattered by an enormous roar. Countless jagged, lightning-like flashes pierced the blackness, illuminating the bleak Cologne Plain. Up and down the American bank for 25 miles, 2,037 big guns, which had been firing only occasional rounds, sent thousands of shells bursting in ripples of orange flame into German positions on the far shore of the swift-flowing stream. The heaviest ground bombardment by an American army in the war had kicked off Operation Grenade.

Bright red chains of tracer bullets streaked across the leaden waters. Rockets swished, shells whined, flat-trajectory rounds from tanks and tank destroyers whizzed past just overhead. Mortar projectiles popped as they flew out of barrels. A curtain of steel and explosives was being lowered onto the Feldgrau on the far side of the Roer. The crescendo was ear-splitting. Soon a dull pink glow lit the skies and thick clouds of smoke and dust drifted upward, forming an artificial fog in the moonlight.

The heavy crash of shells violently shook the building near the

Roer bank where Captain Jerry Hooker of the 104th Infantry Division, who would lead his assault company over the river in less than an hour, was holed up in the cellar with several of his staff. The explosions grew louder. On the ground floor of Hooker's house, Private First Class Allen Clawson of Philadelphia was snoring serenely as though it were a peaceful night along the Delaware River back home.

"That guy can sleep through anything!" remarked a comrade.

"Let him alone for a few minutes longer," a young blond lieutenant stated. "No use his sweating it out any longer than he has to."

Clawson snored on.

A lamp on the table in Captain Hooker's command post cast long shadows on the walls and made the youthful faces look old. There was a shuffling of feet and a rustling of gear. No one said anything. It was time to go. Hooker slung a Tommy gun across his shoulder and folded a single blanket and tied it on his back by crisscrossing a rope around his chest. He glanced around the room and casually remarked, "I wish it were darker. It's a lot harder to get shot if the Krauts can't see you." He hung a couple of grenades on his chest strap.

At 3:15 A.M., with the enormous roar of the bombardment still shaking the house, Hooker said calmly to his Timberwolves, "Okay, boys, let's go."

They clattered up the rickety stairs to the first floor and stepped into the night. Up and down the rubble-strewn street ghostlike figures were slipping out of other houses to join in the assault on the Roer. These were the long-suffering infantrymen who would lead the way, carry the ultimate burden—and absorb 90 percent of the casualties.

Nearby, Private First Class Morton R. Lieberman of the 329th Combat Engineers was crouched nervously in the command post of another assault infantry company as H-hour edged ever closer. The CP was in an old creamery opposite Düren. Lieberman and his comrades had eaten "breakfast" at 9:00 P.M. the previous day, then had moved forward to the Timberwolf company headquarters near the waters of the Roer. The engineers would share the perils with the infantry in equal measure; it would be their job to ferry the riflemen and machine-gunners to the far side.

Ten minutes before H hour, Private Lieberman and a squad of

Timberwolf infantrymen began carrying a bulky, heavy assault boat to the river's edge. It was no easy job. The men were loaded with equipment, and, despite the chill, the struggling assault troops were perspiring freely. Lieberman's boat seemed to get heavier as the little group moved forward, foot by foot, through the oozing mud along the bank of the Roer. On either side could be seen the shadowy silhouettes of other groups lugging boats to the water.

The first waves of Timberwolves jumped off precisely on schedule—3:30 A.M. One engineer was in the bow of each boat setting the stroke, and the infantrymen were paddling furiously against the rush of the current. Lieberman offered up silent thanks for an unexpected development—a heavy haze of smoke and fog, courtesy of the unprecedented Allied bombardment, hovered over the stream, concealing the vulnerable boats.

The 14 Timberwolves in Lieberman's craft stroked and struggled. Their breathing grew labored, and several times the assault boat was nearly swamped. Halfway across the still swollen, sullen Roer, machine-gun rounds streamed over their heads. Then artillery shells and mortar rounds exploded in the water and on the banks. Screams pierced the din on either side as enemy shells and bullets found their marks. After what seemed an eternity, Lieberman's boat scraped the bottom on the far shore. With bullets sizzling past them, the Timberwolf infantrymen scrambled up onto the muddy bank and, crouching low, moved inland.

Hard on the heels of the assault infantrymen, Sergeant William Eckhoff of the 329th Medical Battalion was driving a "weasel" (a small, all-purpose tracked vehicle) toward the Roer. Along with medical supplies, the weasel carried two squads of litter bearers and two technicians. As they neared the river the ground became soft and marshy, and the sturdy weasel was barely able to slog through the flooded terrain. They came upon a branch of the Roer that was too deep and swift to cross with a weasel. So they located an infantry assault boat and paddled to the other side, as occasional German shells exploded around them.

Back on the Allied side were 200 yards of flooded lowland to wade through, ankle and knee-deep, before Sergeant Eckhoff and his medics reached the main stream. Again they were able to locate an assault boat, and crossed the river. An aid station was hurriedly set up in a factory 300 yards from the Roer. Within minutes, infantry combat medics were bringing in casualties, who were administered plasma, morphine, and whatever first aid was required. Litter bearers then took over and carried the stricken Timberwolves from the factory in Düren to the river, where other medics ferried the wounded men back over the stream and to the basement of a gutted church in Merken.

The litter bearers quickly became fatigued from toting their heavy burdens through the deep, slushy mud. Sergeant Eckhoff spotted a group of captured Germans being brought back from the front line and quickly commandeered the services of the enemy soldiers. Under the watchful eye of infantry guards, the prisoners carried scores of litters back to the Roer bank opposite Merken.

Inside the dimly lighted aid station in the Merken church basement, a beehive of activity erupted as the assault on the Roer got underway. Wounded Timberwolves were brought in on stretchers, their uniforms bloody, wet, and caked with mud, as, outside, shells crashed around the old church. Most looked ashen and were shivering from the cold as well as from the effects of their wounds. Some Timberwolves, mainly those with head wounds, were carried in unconscious, and only their deep, labored breathing identified them as still being alive. Others arrived dying, there being no hope for them apart from injections to ease the pain of their last minutes.

Miles to the north of the 104th Infantry Division crossing at Düren, in Simpson's Ninth Army sector, Private First Class Leroy Carver was trudging toward the moonlit Roer through the battered town of Linnich with his comrades in the 1st Battalion, 334th Regiment, of Alexander Bolling's 84th Infantry Division. Helping carry a heavy boat to the water's edge, Carver, like many others in the assault, was thinking of home. Would he ever see it again? "There's too damned many bullets flying around and two many shells crashing here to suit me," he reflected.

Platoon Sergeant George H. Hale of Carver's company had a curious thought as he marched through the mud to the river: "This bombardment is so heavy that we're in danger of being shocked ourselves!" He could not understand how the enemy on the far side of the river could hold up under such a massive pounding.[1]

Along the 84th Infantry Division sector, the infantrymen and engineers slithered to the water's edge. At precisely 3:30 the artillery and mortar barrage shifted several hundred yards back on the far side of the Roer, and the first wave of 35 boats carrying two companies headed for the enemy-held bank. During the 10-minute struggle to reach the other side, a few German mortar and artillery shells exploded among the assault boats, and an enemy machine gun picked off three men in one of the craft.

On the far bank, Lieutenant Richard Hawkins leaped onto the marshy ground, called out in a loud stage whisper, "Okay, let's go!" and moved forward. Hawkins could not believe his good luck. In what had been considered a "suicide assault," the lieutenant had made it over and was still in one piece. Casualties among the "Railsplitters," as the 84th Division was known, had been negligible.

Scrambling out of the boats, the men dropped their rubber life belts and headed for a railroad track some 400 yards inland. Lieutenant Hawkins, Platoon Sergeant Harry L. Peiffer, and other men in A Company dreaded this dash over open ground. Combat experience had convinced the men that the Germans had spent the preceding months sowing thick mine fields in the mushy ground stretching back from the river. Infantrymen of A Company, expecting to be blown to powder with each step, reached the railroad track without a single casualty. There they slumped down with relief.

Only later would Lieutenant Hawkins discover that he and his men had walked right through a field full of stake mines, but that all the trip wires attached to the mines had been cut by American artillery and mortar shells. Not a single mine had exploded, even though a few of the Railsplitters had tripped and fallen over the mines themselves.

Upstream (north) of the 84th Infantry Division, a patrol of Frank Keating's 102nd Infantry Division had slipped over the river 30 minutes before H-hour. After a brief firefight, the patrol knocked out four German machine-gun nests with rifles and grenades. When assault elements of the 407th Infantry Regiment paddled across at this site minutes later, they were met only by meager small-arms fire. Near hits by mortar and artillery shells upset several assault boats, but the men, floundering in the paralyzingly cold and rapid current, were saved from drowning by rubber life vests.

Around Jülich, General Gerhardt's 29th Infantry Division was unable to build bridges because the flood-swollen Roer was 400 yards wide in that sector. Undaunted, the assault and follow-up waves crossed the entire width by paddling small boats and in LCVPs (landing craft-vehicle, personnel), something of a minor naval amphibious operation.

North of Jülich, Leland Hobbs's 30th (Old Hickory) Infantry Division faced great difficulties. At only two sites was the Roer narrow enough for crossing. Near the village of Schophoven, a 25-man patrol sneaked across at 1:45 A.M., more than an hour before the start of the bombardment and 105 minutes before H-hour. The patrol deployed and furnished a protective screen as engineers followed in assault boats, dragging behind them prefabricated duckboard bridges. Forty-five minutes before H-hour, just as the enormous American bombardment erupted, a battalion of Old Hickories paddled across the treacherous Roer, the first sizable American force to reach the German-held side. By the time the barrage shifted inland at H-hour, the remainder of the assault regiment was moving across on the duckboard bridges.

By dawn, all up and down the 25-mile stretch of the Roer River, the last major barrier before the Rhine, American troops were on the German side in strength. Operation Grenade was off to an encouraging start. But at their headquarters, General Simpson of Ninth Army and General Collins of the First Army's VII Corps were feeling uneasy. German artillery had knocked out many foot and vehicular bridges, and most American units were fighting on the far side of the swollen Roer without armor or anti-tank guns. The element of surprise had vanished, and the Germans were showing signs of organizing counterattacks to wipe out the bridgehead.

At 2:55 P.M. a grim-faced Collins hopped out of his muddy jeep and strode briskly into the command post of the 413th Infantry Regiment opposite Düren. The general was clad in the faded trenchcoat that had become a personal trademark in countless battles across the face of northern Europe. Collins, a bold and aggressive combat leader, was concerned. The crossing by Collins's corps had been the most difficult. In his sector on both sides of Düren the banks were higher, the current swifter. Inside gutted Düren, reports were being sent back that tough, first-rate German soldiers, including paratroopers, were defending the town house by house. A bridge capable of carrying tanks to the far shore still did not exist in the entire VII Corps sector.

General Collins was hurriedly briefed on the situation on the far shore. He then stressed the importance of seizing a large insane asylum in Düren that was expected to be a focal point of German resistance. "When are you going to take it?" Collins inquired of Colonel Welcome P. Waltz, commander of the 413th Regiment.[2]

Before Waltz could reply, a radio signal came in from Lieutenant Colonel William M. Summers of Tulsa, Oklahoma, commander of the 3rd Battalion: "We are on objective." The objective was the asylum complex, known to the Germans as "the provincial health and welfare institute." Patients at the facility had previously been evacuated or wandered off in the confusion on the German side of the Roer.

Colonel Summers's K Company, led by Captain Hayden Bower of Hollywood, California, had been involved in bitter house-to-house fighting in northern Düren since reaching the far shore of the Roer shortly after H-hour that morning. He and his men had fought through six floors of a large paper factory, flushing out Germans at each level. Reaching the roof, the Americans had captured four enemy artillery observers huddled in a concrete pillbox.

It was Captain Bower and his men who seized the insane asylum. Prior to jumping off in the river assault, General Terry Allen, leader of the Timberwolves, had ordered the dominating asylum to be captured by 3:00 P.M. Bower radioed its seizure at 2:57—three minutes under the deadline.

Like nearly every other structure in gutted Düren, the asylum had been pounded for many weeks by bombers and artillery, and its upper floors now consisted only of brick sieves, although a labyrinth of cellar rooms was filled with comfortable divans and other evidence that the Germans had, as suspected, used the implied immunity of the insane asylum to locate a headquarters there.

Meanwhile, Lieutenant Colonel Summers's I Company, advancing on Captain Bower's right, was involved in a savage fight in a block of apartments across from Düren's electrical power plant. I Company men cleaned out the apartments and moved on to flush the enemy from the power plant. The Germans, however, slipped back into the large apartment complex. Again a fierce, close-quarter fight erupted in the apartments, room by room, mainly with grenades, Tommy guns, and rifles. It took several hours to wipe out or capture the last German.

Late in the afternoon of February 23, after 13 hours of heavy fighting, about half of Düren had been cleared. A stench of death hung over the rubble of the town: long-neglected bodies of soldiers and civilians alike were buried under tons of debris. What the air corps had not done, the artillery and mortars had. Here and there fires were burning in the skeleton of Düren. Not a single civilian was in sight, although some who had insisted on remaining were later found huddled in basements.

Standing in the main square of the Rhineland city, two drained American soldiers were gazing at an enormous bronze statue. The statue was the only thing still standing in the square.

"Who's that old bastard?" Corporal Walter Makara inquired idly.

"Goddamned if I know," replied Corporal James "Red" Groves. "Some Kraut bigwig, I guess."[3]

The statue was that of Otto von Bismarck, Germany's famed Iron Chancellor of the previous century, who appeared to be surveying the devastation surrounding him.

At the First Army press camp west of the Roer River that night, war correspondent Graham Miller typed a story for dispatch to his newspaper in the States:

Düren is the most completely smashed town I have seen in this war. It is gashed, splintered and torn, with here and there an acre or so of nothing but powdered rubble that grits between your teeth as the cold wind blows it about. . . . There's not a house, not a tree, not a blade of grass.[4]

It was 9:15 P.M. when elements of Bolling's 84th Infantry Division drove the last German out of the village of town, about two miles northeast of their Roer crossing site. Marching and fighting for more than 30 hours, the Railsplitters were near exhaustion. In the gutted town, they flopped down to snatch a few hours of sleep.

At 10:30 P.M. their slumber was interrupted by the roar of powerful tank motors to the rear of Baal. "Probably our tin cans got over the Roer and have caught up with us," a drowsy rifleman observed. Moments later three German Panther tanks and a force of grenadiers (infantrymen) burst into Baal from behind the village. The panzers raked a company CP with point-blank cannon and machine-gun fire. Scrambling for cover, GIs were amazed—the enemy tanks had their headlights blazing.

Pandemonium erupted in Baal. Two Railsplitters were herding 30 German prisoners back down the main street when they ran into the panzers barreling through town. In the darkness and confusion of the wild firing that emerged, the POWs scattered and escaped.

Along a railroad embankment toward the front of the village, several German machine guns began firing streams of tracers back into town. As the enemy tanks rolled on through the debris-strewn streets, a few Railsplitters edged out of the shadows, waving and yelling, trying to warn the crews of the tracked vehicles, which they mistakenly thought were friendlies, of the German machine guns at the railroad. The startled Americans flopped to the ground as their warnings were greeted with a fusillade of machine-gun fire from the panzers.

Shooting up the buildings on either side of the street with bursts from their automatic weapons, the German tanks clanked through Baal and disappeared into the darkness at the front of the village, as frustrated Railsplitters peppered the rear of the armored vehicles with ineffectual rifle and machine-gun fire.

About two hours later, Lieutenant William Nelson, a rifle platoon leader, was huddled with his men in a battered old building in Baal, defending an underpass and a junction on the road leading to Granterath. Suddenly, intense firing erupted in the darkness. Bullets beat a steady tattoo on Nelson's building. Only later would the lieutenant learn that a hodge-podge force of Germans, supported by three armored assault guns, had launched a counterattack to drive the Americans out of Baal.

Lieutenant Nelson and his men fired back into the darkness. The Germans were difficult to locate. Like a circled wagon train in the Old American West, Nelson's platoon fired in all directions with machine guns, rifles, and Browning automatic rifles, and kept the attackers from storming the cornered Americans. But now a new danger emerged—Nelson and his men were nearly out of ammunition. The Germans, cloaked in darkness, seemed to be edging ever closer to the structure.

The American situation was critical. Nelson arrived at an anguishing decision: he would call for artillery on his own position, hoping that the Germans out in the open would get the worst end of the bombardment. The platoon had no radio, so Nelson sent two runners to the company CP. They never got there.

In the meantime, Lieutenant Nelson's company commander had learned that Germans were swarming around the underpass near Nelson's building, so on his own he called for artillery fire. As the trapped Americans awaited a rush by the enemy grenadiers, they heard the eerie swishing sound of incoming shells, and seconds later explosions ripped the terrain around the underpass. For an hour the Railsplitters huddled anxiously in the darkness of their building as the area around them was heavily shelled by American artillery.

Eventually, Nelson and his men heard German voices at the underpass crying out for the terrifying barrage to cease. A German-speaking soldier in Nelson's platoon shouted for the enemy around the underpass to surrender. Individually and in pairs, the battered Feldgrau, with hands raised, trotted toward the American platoon's building. Some dragged or carried wounded comrades. Finally 25 Germans were captives in the cellar of the American-held building. Almost incoherent from the savage shelling, the grenadiers claimed that they were all that were left of the more than 200 Germans who had jumped off from Granterath to drive the Americans out of Baal.

In and around Baal a series of sharp firefights and hand-to-hand battles continued on through the night. The first streaks of daylight revealed German and American bodies strewn about the streets, but the Railsplitters clung desperately to the gutted, insignificant little village.

At midnight on February 23—D-day for Grenade—Generals Bill Simpson and Joe Collins and other top American commanders were beginning to breathe more freely. The bloodbath for their assault troops, which some feared might take place in the difficult crossing of the flood-swollen Roer, had not materialized. Ninth Army and First Army's VII Corps had lost large amounts of equipment, floating bridges, and assault boats. But these could be immediately replaced from enormous stocks. In the crossing, Ninth Army had had 92 men killed, 913 wounded, and 61 missing, while Collins's corps had had 66 men killed, 281 wounded, and 35 missing. These human losses, too, could be rapidly replaced.

6

Hitler Rejects a Pullback Over the Rhine

On the chilly morning of February 24, weary Americans along the Roer peered into the bright sky at a curious phenomenon: black-painted Luftwaffe warplanes were heading for the foot and vehicular bridges that were the lifelines of the assault troops on the east side of the Roer. Seeing German aircraft in the daytime had been a rarity since the Allies invaded Normandy the previous June. Now here were groups of four or six, instead of the occasional darting raids by a single Luftwaffe daylight intruder.

In Düren, Platoon Sergeant Robert Murphy of New Orleans, Louisiana, and comrades in the 87th Mortar Battalion were puzzled as they gazed into the sky. These German aircraft looked unlike others they had seen; they did not have propellers.

"What in the hell is holding them up in the air?" a mortarman called out.

"Mobile skyhooks," replied another.[1]

Large numbers of Thunderbolts and Mustangs were patrolling, and several squadrons took out after the Luftwaffe planes. But the sleek German warplanes left their pursuers behind as though the American fighters were standing still.

As the Allies soon discovered, these were the new and notorious (to the Allied high command) Messerschmitt-262 jet fighter-bombers, revolutionary aircraft that were faster and more maneuverable than Allied propeller-driven planes. A few of the German jets had swooped over the Roer on reconnaissance missions in recent weeks, but rarely, and only individually. But with Eisenhower's forces across the final natural barrier to the Rhine, Adolf Hitler was committing one of his "secret weapons" in increasingly large numbers.

There was a chilling aspect to the Messerschmitt-262 to those on the ground. The ME-262s would approach a target from high in the sky, then idle their engines and glide down in silent dives that enabled them to loose their bombs or strafe forward American ground units without warning. Soon the constant threat of the diabolical German jets, which struck literally from out of the blue, began to gnaw at the nerves of the men on the Cologne Plain.

On a typical flat piece of ground east of the Roer, a battery of the 555th Antiaircraft Artillery Battalion was furiously pumping shells into the sky, aiming at a flight of Messerschmitt-262s. Commanded by Captain Fred D. Waters of Chicago, the battery was having a difficult time finding the range, although puffs of smoke erupted continually around the jets. "Keee-ryst, those goddamned things are fast!" complained Corporal Joe Idasek of Cicero, Illinois, whose job was to set the range and keep the bursts ahead of the planes.

The antiaircraft gunners had quickly learned the German jets' silhouette and could spot it far off, even by moonlight. But spotting it and hitting it were different matters. This flight of ME-262s hurried over, and Captain Waters's guns ceased firing. Moments later the jets' bombs could be seen cascading earthward, followed by the *kaaarrump* of the exploding ordnance. The jets were going after the Roer bridges.

Yet the German jets were not immune to ground fire. Early in the afternoon of February 24, D-plus-one for Grenade, two ME-262s ame in to bomb and strafe bridge sites north of Linnich. Gunners of Battery C of the 557th Antiaircraft Artillery Battalion threw up a curtain of fire, hitting the first jet as it was making its bombing run. It crashed with an enormous display of orange flame and black smoke. The GI gunners cheered their first conquest of the "unhittable" *Wunderwaffe*.

Undaunted by the loss of his wingman, the pilot of the second jet dove on a Roer bridge and released his bomb. The marksmen of Battery C actually hit the bomb in midair, detonating it, then turned their weapons on the aircraft. The Luftwaffe pilot put his ME-262 into a nearly vertical climb, but the jet was struck by a shell and exploded in a fiery ball.

It was nearing midnight on February 24 as Timberwolves of Allen's 104th Infantry Division were silently filing into thick woods outside Arnoldsweiler. These were men of Captain Walter Leigon's F Company, which had been given the mission of seizing a massive medieval castle. Reaching a clearing, Leigon's men could discern the structure's silhouette against the dark sky a hundred yards ahead.

Schloss (castle) Rath was a red brick building with turrets on each corner and a 15-yard moat surrounding it with water to a depth of five feet. While one platoon remained at the edge of the woods in reserve, another platoon assaulted the main gate in an effort to gain entrance by surprise. A German sentry at the wire gates in front of the moat was dispatched silently by means of a trench knife in the hands of the platoon's lead scout. On a signal from the scout, the remainder of the platoon charged the gate. On entering the castle grounds the Americans were raked by fire from four halftracks mounting flak guns and two self-propelled guns inside the enclosure. The platoon hastily took cover behind a stable, where it was pinned down.

In the meantime, another platoon, led by Lieutenant Calvin Walker of Denver, Colorado, had edged up behind the castle and slipped over the moat and onto the grounds before being discovered. Heavy hand-to-hand fighting broke out in the darkness, and Walker's men seized all the buildings within the grounds except the castle itself. Hurriedly assembling his force, Walker led them in a charge against a stone archway that led into the building, as BAR men peppered the windows to prevent the Germans from firing out of the openings.

A German with a Schmeisser machine pistol popped out from behind the archway and cut down Lieutenant Walker's platoon sergeant. Another American quickly shot the German and the rest of the platoon stormed into the *schloss*. Inside the cavernous old structure a wild shootout erupted in the near-blackness. The sounds of Tommy guns and Schmeisser machine pistols echoed up and down the long marble halls as the GIs chased the Germans from room to room and finally cornered the survivors in the cellar.

"*Komm heraus, Schweinehunde!* [Come out, you sons of bitches!]" a leather-lunged American shouted down the dark stone steps leading into the lower chamber. The call was greeted with silence.

"Okay, here's a present for you bastards," a GI yelled as he tossed a concussion grenade into the cellar. A loud explosion was followed by thick clouds of black smoke floating up the stairs. Speaking in perfect English, a voice in the lower chamber called out, "We surrender! We surrender!"

Led by Lieutenant Walker and with guns at the ready, the Americans cautiously edged down the steps—and discovered that they had bagged an entire battalion staff of the 10th Panzer Grenadier Regiment, including a lieutenant colonel, six other officers, and 100 men. In one corner of the basement were 12 men wearing wide grins on their faces—Timberwolves who had been captured by the Germans earlier in the day in fighting at Arnoldsweiler, a few miles west of Schloss Rath.

With the entire castle complex in their hands, Captain Leigon's F Company men settled down for a respite. It was short-lived. A German force that had been kicked out of Arnoldsweiler was headed for Schloss Rath, not knowing it now had American tenants. On finding that the medieval castle had changed hands, the German force opened fire with automatic weapons, rifles, and six self-propelled guns.

Lieutenant Walker, leaving only three men to guard the 107 enemy POWs, rapidly deployed his men to meet the German attack on the castle. As the din of battle indicated to the prisoners that their comrades would soon overrun the building and free them, an Englishpeaking German lieutenant snarled at Walker, "Now you're going to et what you've got coming!" Angered, Walker raised his carbine, hen thought better of it. With a large force of Germans closing in, here was a distinct possibility that the tables indeed might be turned, o Walker cursed the enemy officer and departed.

Reinforced by the 12 rescued Timberwolves, Walker and his men raced for the enemy charge. That charge never came. The F Company platoon that had been positioned at the edge of the woods when the other two platoons assaulted the castle spotted the 200 Germans and six SPs (self-propelled guns) advancing toward Schloss Rath. Platoon Sergeant Oscar S. Lycksel, who had taken command when the platoon leader was killed, ordered his men to hold their fire until he gave the order.

Moving into an open field, the attacking enemy force was unaware of the Americans just inside the woods to one side. Suddenly, Lycksel and his men cut loose with all of their firepower, bringing down the exposed Germans as though their ranks had been struck by a gigantic scythe. Within minutes the field was littered with bodies in field gray uniforms. A handful who survived the withering fusillade fled.

The battle for Schloss Rath was over. "I think this goddamned Kraut castle needs a *W* in front of that *R* in Rath," mused an exhausted Calvin Walker to his men. Elsewhere on the castle grounds, F Company commander Captain Leigon was toting up results for the night of heavy fighting: 396 prisoners plus 175 Germans killed, and scores more of the enemy wounded. "Not a bad night's work for one company," Leigon observed.[2]

Along Simpson's Ninth Army front on February 24, Major General Leonard Baade's 35th (Santa Fe) Infantry Division, attacking over the narrow Wurm River, ran into heavy resistance. F Company of the 320th Infantry Regiment reached a field saturated with Schu mines. Under Captain James M. Watkins, the company tried to fight its way forward. One soldier in the minefield set off an explosion; his leg and foot were blown from his body and landed in a road 90 feet away. Howling in agony, the man bounded to one side on his stump and knee, setting off another mine, which killed him.

In the same minefield, another of Watkins's men had one leg blown off and the other so badly shattered that it hung crazily, the foot pointing in the wrong direction. Alone in the center of the minefield, the soldier managed to apply a tourniquet to the stump to halt the gushing of blood, and bandaged his other wounds, although very crudely. All the while, German machine-gunners periodically raked the field with fire.

Nearby was the body of a comrade, killed by a mine. The wounded soldier slithered over to the corpse and scooped out a shallow excavation behind the lifeless body, which offered some protection against enemy bullets. Never losing consciousness, the Santa Fe Division soldier lay in the cold and rain far into the night while engineers worked their way out to him. When the rescuers reached him the wounded man was in unaccountably good spirits, asking them, "What brings you fellows out on a night like this?"

That same day, Captain Orval E. Faubus of Arkansas, a staff officer in the 35th Division's 320th Infantry, was jeeping forward with his driver, Private First Class Henry L. Lanier of Ferreston, Texas. Faubus reflected on the term "smoke of battle" as ahead tall black plumes spiraled into the air where German artillery had found its mark.

Approaching a road junction, Faubus noted that a truck was burning fiercely in the center of the intersection, and that German shells were pounding the crossroads. A GI truck driver leaned out of his cab and shouted at the passing jeep, "Don't go up there! Don't go up there!" At Captain Faubus's urging, Lanier pushed down hard on the accelerator. Up ahead, a shell struck directly in the center of the intersection, creating a thick cloud of smoke. Reaching the crossroads, Lanier slowed to make a sharp right turn, edging past the burning truck, and the two men ducked as several shells came screaming into the junction, neatly bracketing the jeep and sending shrapnel fluttering overhead.

A short distance down the road, Lanier, his heart still pounding heavily, asked, "Captain, what did that guy back there on the other side of the crossroads shout at us?"

Faubus replied, "He said, 'Don't go up there!'"

Swallowing hard, Lanier said, "That's what I *thought* he said!"[3]

* * *

By the end of the first two days of Grenade, the Americans had captured some 4,000 Germans along the flaming Cologne Plain. The enemy prisoners were a mixed bag: first-rate troops in their early twenties and late teens, railroad battalions, boys of 15 and 16, Luftwaffe and ground service personnel, and members of the Volkssturm, some of whom were boys of 12 and 13. All over Germany, Hitler had hustled Volkssturm recruits into the fighting near their homes on the theory that they would willingly die defending their own turf. The Volkssturmers quickly discovered that each time they put up a fight, the Allied juggernaut reduced their towns to rubble. They found that the surest way to spare their own lives, homes, and property was to promptly surrender. This capitulation was infectious in the closing weeks of the war: quite often, regular members of the German army followed the example of the ersatz soldiers.

As Allied forces drove forward, nervous German generals began to show renewed interest in the Rhine bridges. Defending the spans was the responsibility of the *Wehrkreise*, the administrative regions into which the Third Reich had been divided. In command of the *Wehrkreise* was Heinrich Himmler, the one-time chicken farmer who had become Hitler's SS and Gestapo chief. The procedure was for the *Wehrkreise* to relinquish control of the bridge defenses as soon as the Allies neared a specific locale; at that point the regular fighting troops would take over. But bureaucracies die hard. As late as February 25, with Eisenhower's forces only 20 miles from the Rhine at some points, the *Wehrkreise* bureaucrats clung to their fiefdoms and refused to cede their authority over the crucial bridges.

For weeks Walther Model, commander of Army Group B, had had his chief engineer, Lieutenant General Jankowski, feverishly developing detailed plans for protecting the bridges and ferries. Jankowski's task was one of deep frustration. He was forced to contend with jealousies on all sides, with *Wehrkreise* bureaucrats and combat leaders reluctant to furnish information for fear of helping their rivals gain influence over the Rhine bridges. Jankowski was never certain what channels of command to go through, and Model—who was deeply involved in trying to patch up the lines recently breeched by the Allies—was of little help.

Meanwhile, on the northern sector, the Canadian First Army under Henry Crerar had been slugging its way forward for two weeks in Operation Veritable. Paced by Horrocks's British XXX Corps, the Canadian army had battled floods, knee-deep mud, heavy rains, fortified towns and villages, and resolute German paratroopers in seizing its two primary objectives, Goch and Cleve (the latter town was known to history as the home of Anne of Cleves, fourth wife of Henry VIII of Great Britain).

The seizure of these two towns in some of the fiercest hand-to-hand fighting of the war brought a measure of relief to Field Marshal Montgomery, for Cleve and Goch were reputed to be the last major enemy strongholds before the Rhine River. But each German town and village after that proved to be an equally tough nut to crack. There would be no startling breakthrough on the flooded Canadian and British sector to Hitler's historic moat. Some 10 German divisions, well armed and most in fortified positions, were crammed into the narrow strip of terrain between Montgomery's attacking forces and the Rhine, and they gave every indication of fighting to the bitter end.

South of Canadian and British forces on the morning of February 25, the third day after the launching of Grenade, "Texas Bill" Simpson was encouraged by the situation in the Ninth Army sector. Despite regular attacks by German jet aircraft and pounding by artillery, Ninth Army engineers had thrown up more than 20 foot- and vehicular bridges, permitting a steady flow of tanks, supplies, and reinforcements over the Roer. The terrain in front of Simpson, all the way to the Rhine, was a tanker's dream—flat country generally devoid of trees, with an ideal hard-surface road network.

Simpson's infantry had done its job by poking holes through the German lines. Now the Ninth Army commander cut loose his armor for the dash to the Rhine. The combat troops of Major General Lunceford E. Oliver's 5th Armored and Brigadier General Isaac D. White's 2nd (Hell on Wheels) Armored Divisions plunged ahead. High above in a Cub aircraft, correspondent Sidney Olson of *Time* magazine was viewing the 2nd Armored's advance. To Olson, White's tanks crawling across the barren Cologne Plain resembled huge black beetles, aided in their advance through cabbage fields by swarms of Mustang and Thunderbolt fighter-bombers pouncing on German strongpoints at the front.

At his headquarters in Bonn on the afternoon of February 25, Field Marshal von Rundstedt was growing increasingly alarmed. The Allied strategy was now clear to the aging commander: Veritable, driving to the southeast, and Grenade, advancing to the northeast, were to link up west of the Rhine. If they should succeed in this, at least two German armies, which contained some of von Rundstedt's best troops, would be cut off and destroyed.

But von Rundstedt had worries other than the looming Wehrmacht disaster in the north—chief among them being the unpredictable General George Patton in the Eifel far to the south. German leaders had long considered the audacious Patton the Allies' "most dangerous" battle commander. "Wherever Patton is, that is where things are going to happen," Wehrmacht generals reminded each other. And von Rundstedt was well aware that Collins's VII Corps was heading hell-bent for Cologne, Germany's fourth-largest city.

Von Rundstedt fired off an urgent signal to Hitler asking for "new directives." Unless there was a general pullback across the broad Rhine, the entire western front could collapse, declared the commander in chief, west.[4]

The appeal was ignored by the Führer.

That same afternoon orders were issued for elements of Terry Allen's 104th Infantry Division to attack and seize two fortress towns on the Cologne Plain, Morschenich and Golzheim. General Collins planned to unleash Major General Maurice Rose's 3rd Armored Division in a dash for the Rhine the following day, and he did not want the armor held up by stubborn German opposition in those towns. The Timberwolves were given a specific time to clear Morschenich and Golzheim—6:00 the next morning. This meant mounting a night assault, one of the most difficult military operations, but one in which the 104th Division was considered expert.

Advancing through a forested area, guided by the muted light of the moon and by the Düren-Cologne railroad track, elements of the 104th Division rushed Morschenich, overran three enemy tanks, and by 5:45 A.M.—15 minutes in advance of the deadline—radioed back that they had secured the town and captured 315 prisoners. Earlier that morning, other Timberwolves, many mounted piggy-back on tanks, charged Golzheim and by 2:00 A.M. had captured the town only 14 miles from the key objective of Cologne.

After Golzheim was secured, Sergeant James T. Sobansky, of Washington, Pennsylvania, and a few of his men entered a shell-torn building in search of snipers—or souvenirs. They were not prepared for what they found. They heard a rustling in one room, kicked the door open with weapons at the ready, and gaped in astonishment. Inside were four rather attractive women in their mid-20s, each clad only in panties despite the cold, and four men wearing either long johns or nothing. "I guess we interrupted the hottest lovers' rendezvous in Europe," Sergeant Sobansky remarked dryly, lowering his weapon. "And with all those bullets and shells flying around them . . ."

Other Timberwolves swarmed into the house to have a look. Lieutenant Stanley R. Blunck of Oakland, California, determined that the men were Polish slave laborers left behind when the Germans in Golzheim fled. The nationality of the scantily clad women was unknown. A Timberwolf report to higher headquarters stated: "Golzheim seized. 227 Germans and 8 lovers captured."[5]

On the evening of February 27, General Collins issued orders for the final assault on Cologne, fabled in German folklore as the Queen City of the Rhine. The Germans in front of Collins's troops were in disarray; he intended to keep up the unrelenting pressure. That day his 3rd Armored, 8th Infantry, and 104th Infantry divisions had plunged forward, reaching the last remaining natural obstacle before Cologne and the Rhine—the Erft Canal complex. This barrier consisted of three canals, each several feet deep and ranging from 12 to 30 feet in width. Cologne was only 10 miles away.

General Maurice Rose, the stern-faced, stolid son of a Denver

bbi, called in two of his 3rd Armored Division task force command-
s, Lieutenant Colonels Walter B. Richardson and Samuel Hogan.
oth men were Texans and dashing combat leaders, and had an in-
nse desire to outdo each other on the battlefield. Rose told the two
eutenant colonels that the 3rd Armored, along with other divisions
VII Corps, would attack over the Erft Canal, and added, "I'll give
case of Scotch to the first one of you to cross the Erft." Neither task
rce leader needed the liquid incentive. Beating the other over the
ater barricade would be reward enough.[6]

Dusk had turned into the blackness of night on the Cologne Plain
hen gunners of the 991st Armored Field Artillery Battalion loaded
ells into their 155-millimeter Long Toms and set the range at
5,100 yards. At 6:28 P.M. (February 27) the huge weapons roared
nd the projectiles catapulted eastward into the dark. Minutes later
ne shells exploded in ancient Cologne. For the first time in the war,
Germany's fourth-largest city was under ground fire.

In the meantime, while the German army was trying to stem Gen-
ral George Patton's full-blooded "probing attacks" in the Eifel, the
ong-festering animosity between Field Marshal Walther Model and
General Brandenberger, commander of Seventh Army, came to a
ead. At a tension-racked conference of corps and division leaders,
Model severely castigated Brandenberger, accusing him of incompe-
ence and cowardice. He concluded his tirade by sacking Branden-
erger on the spot and replacing him with Lieutenant General Hans
elber, commander of XIII Corps, while Felber looked on in embar-
assed silence.

Brandenberger promptly left the building housing Seventh Army
eadquarters after his public humiliation by the sharp-tongued
Model. Minutes later an American fighter-bomber dived and dropped
500-pound explosive directly onto the structure. Whatever he
hought about the disgrace of being fired in front of his generals for
incompetence and cowardice" (charges other German leaders
randed as false), Brandenberger's life was probably saved by the ac-
ion.

The loud explosion rocked the Seventh Army headquarters build-
ng, showering those inside with broken timbers and masonry. Model

escaped uninjured but shaken. Felber received a minor head wound. The blast stripped the clothes off the Seventh Army chief of staff, but he received only a cut on the head. Several staff officers were killed or seriously wounded.

With Hans Felber's promotion to head Seventh Army, Model also appointed Lieutenant General Ralf Count von Oriola to command XIII Corps. Faced with Patton's "aggressive defense" attacks, Oriola promptly made the same request to the new army commander, Felber, that Felber had made to Brandenberger—permission to withdraw to more defensible positions. The new Seventh Army commander, as a consequence of Hitler's "no retreat" writ, found himself having to deny the precise request that he himself had made a few days before as a corps commander.

Knowing that the zealous Model was on guard at all times to thwart any "defeatist" action, General Felber had to resort to a strategem to circumvent the will of the Führer, one that could have resulted in his facing a firing squad. Working through his chief of staff, Major General Christoff Georg Count von Gersdorff, Felber advised his subordinate commanders that they would receive *two* versions of orders. One would direct the unit to stand firm and was to be filed in official records for Hitler's consumption. The second would give the order that General Felber actually intended to be followed, and was to be promptly read and burned. To justify withdrawals, Felber's operational orders to higher headquarters would be falsified.

7

Vaulting the Erft Canal

In Berlin on the night of February 27, Adolf Hitler was raging about Field Marshal von Rundstedt's urgent appeal to pull back over the Rhine. Surrounded by functionaries in the high command, Hitler declared, "We must cure him [von Rundstedt] of the idea of retreating. These people [German battle leaders] just don't have any vision. It would only mean moving the catastrophe from one place to another. . . ."

Still, the World War I infantry corporal seemed to have gnawing concerns as to what really was taking place on the western front. He exclaimed that he didn't trust field reports from his commanders. "We have to get a couple of officers down there [west of the Rhine]—even if they have only one leg or one arm—officers who are good men, whom we can send down there to get a clear picture."

In his desperation to halt the Allied steamrollers in the west and east, Hitler suggested mustering women for front-line combat duty. "So many women who want to shoot are volunteering now that I really think we ought to take them in immediately," he declared. The startling suggestion of sending German women into the front line was received with stone-faced silence.[1]

Earlier that day the battlefield disaster in the west that von Rundstedt had warned Hitler about had become a reality. Elements of Bolling's 84th Infantry Division in Ninth Army jumped off at 7:00 A.M. and, after brushing aside early sporadic resistance, plunged ahead for nine miles. The German lines in front of the Rhine had started to crumble.

In Lenholt, a battery of German 75-millimeter guns was found parked, unguarded, on the main street. Bolling's riflemen leaped off tanks to roust out the gun crews, and found them sound asleep in a nearby building. Disheveled German prisoners began streaming to the rear, some without guards. Americans simply waved them in the direction of POW cages. Without a shot being fired, the tank-infantry task forces pounded through town after town, many of which had scores of white sheets flapping in the wind.

The GI's had a term for a situation like this—a Rat Race.

As the four-mile-long column of Railsplitters drove forward, the 84th Reconnaissance Troop prowled around the flanks on the side roads and lanes, probing villages, woods, and defiles for the enemy. Near Holtum the mechanized cavalry troopers seized eight 88-millimeter guns parked unattended along a road. Seeking the missing enemy gunners, an armored car and several jeeps loaded with riflemen edged into Holtum. Hearing the vehicles approach, two German soldiers stuck their heads out of a window, and were waved outside. Instead of two, 13 Feldgrau emerged with hands raised.

It was only the beginning. Soldiers kept filing out of houses for the next several minutes until the street was jammed with 250 prisoners. A chagrined and confused German major had been eating when the 84th Recon men stormed into town. He admitted he had had no idea that Americans were anywhere near.

During the headlong advance that day, the Railsplitter column had not lost a single tank. Near the village of Steeg at 2:00 P.M. a concealed German anti-tank gun went off and the lead American tank was knocked out. The enemy gun was eliminated, but when the column started up again the new lead tank had traveled only a few yards when it was struck by a high-velocity shell and set afire. It was obvious that an organized defense had been set up at a road junction, and the Railsplitter column ground to a halt.

Brigadier General John H. Church, assistant commander of the 84th Infantry Division, was some three miles back in the lengthy column. He decided to go forward to find out what was holding up the task force. A reconnaissance jeep led the way, followed by Church's jeep and a third one carrying two newspaper correspondents. About a mile forward, several German soldiers dug in along the road began raking the jeeps with rifle fire. The vehicle carrying the reporters spun about and headed for the rear, but the other two jeeps continued onward. A running gun battle erupted between the Railsplitters in the racing jeeps and the Feldgrau strung out along the road.

The firing became so intense that General Church decided prudence was the better part of valor and ordered his driver to turn back. Halfway to their destination on the return trip, the two American jeeps were peppered by rifle and machine-pistol fire from *both* sides of the road. General Church's vehicle was struck and began to reel crazily. The driver, Corporal Kyser Crockett, screamed, "I've lost my arm!" He slumped in his seat, but tried to keep his foot on the accelerator to keep the jeep moving through the enemy gauntlet. The general's aide, Lieutenant Norman D. Dobie, hurriedly leaned over from the back seat and steered the jeep through the curtain of fire, as Church blazed away at both sides of the road with his Colt .45 pistol.

Reaching safety, Lieutenant Dobie noticed for the first time that General Church's face and uniform were saturated with blood. Fragments had struck him on the forehead, knee, and ankle. Corporal Crockett, weak from loss of blood and shock, still had his arm intact, but it had been badly mangled. Despite the insistence of army doctors that he be evacuated, General Church refused to leave the front, and continued in the fight toward the Rhine after a couple of days of partial recuperation.

While General Church was engaged in his running gun battle, two companies of the 84th Division collided with a group of the German 8th Parachute Division barricaded among the shell-torn houses in the village of Berg, west of the major city of München-Gladbach. One of

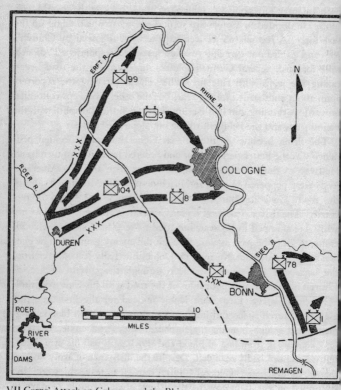

VII Corps' Attack on Cologne and the Rhine

hem, Company G of the 334th Regiment, was crossing 300 yards of open fields to reach the road leading into Berg when four enemy machine guns opened fire. Lieutenant Harold L. Howdieshell, who had just received a battlefield commission, was at the point with two scouts, 25 yards in front of the main body. He shoved the scouts into a ditch and tossed four hand grenades at one chattering machine gun. It fell silent. Howdieshell drew back his arm to pitch a fifth grenade just as another machine gun cut loose. A bullet pierced the young lieutenant's head, killing him instantly.

The second machine gun was spotted firing from a one-story building. Lieutenant Jack F. Schaper slithered forward with bullets passing just overhead and reached a low sugar-beet mound. He peeked around the side of the mound to get a look at the enemy automatic weapon when a slug struck him in the head. Badly wounded, Schaper was evacuated.

The German machine gun was pinning down the entire American company. Captain Charles E. Hiatt, G Company commander, sent a platoon through a patch of woods to outflank the enemy weapon. Reaching the house, the Americans charged and wiped out the machine-gun crew. A German major was slumped over the automatic weapon, its barrel still hot.

Two more machine guns manned by German paratroopers continued to force the Americans to dive for cover. Above the din of the grating machine guns, a Railsplitter called out, "To hell with this shit! Let's rush 'em!" Bayonets fixed, the 84th Division men scrambled to their feet, charged up a low slope, and sprang into a trench system occupied by enemy paratroopers. A wild melee erupted. When it was over, only two German paratroopers, both wounded, remained to be captured.

Captain Hiatt's company promptly jumped off for Eicken, a village 800 yards away. A German machine gun concealed in a haystack raked the leading platoon, pinning it to the ground. Private First Class Max L. Slate crept around to the rear, removed a match from his pocket, and set the haystack ablaze. The enemy crew dashed out to surrender.

As tankers and foot soldiers of General Joe Collins's VII Corps readied to spring across the Erft Canal, just over the horizon ancien Cologne was a ghost city—gutted, blackened, mortally ill. Over 8(percent of it had been destroyed in five years of Allied air bombard- ment. Out of a peacetime population of 780,000, some 80,000 fright- ened men, women, and children remained, living like moles amid the rubble, trembling each time the rumble of Collins's guns to the west reached their ears.

Once the seat of Rhineland culture and possessing in its famous medieval cathedral, the Dom, the finest specimen of decorated Gothic architecture in Germany, Cologne now was a city of despair and soup kitchens and epidemics. Public utilities functioned intermittently, and three of the city's five Rhine bridges had been destroyed by English and American bombers. The famed Hohenzollern Bridge was still standing—but barely. The newest of the five river spans, it was 1,400 feet long, and its west entrance was adjacent to the majestic cathedral with its 515-foot-high twin towers. Also damaged but intact was the Hänge-Brückl Bridge. The northernmost span, the Köln-Mülheim, a 2,300-foot suspension structure, had been destroyed the previous No- vember when a near hit by a bomb detonated demolition charges planted in the vicinity.

Before Cologne had begun to receive special attention from British and American heavy bombers early in the war, its Lindenburg hospi- tal was the largest in western Germany, with a full-time staff of 150 doctors, 200 nursing nuns, and 250 orderlies and servants. It had boasted accommodations for 1,000 student doctors from all over Eu- rope as well as the United States. Now, in Lindenburg hospital's un- derground air raid shelter among the ruins of the dying city, Father Anton Gotzens, a Catholic priest and hospital chaplain, was going from bed to bed, ministering to the remaining 200 patients. They were under the care of 20 physicians and 20 sisters. Each time a flight of American fighter-bombers roared over the shambles of Cologne, the patients, most of whom were elderly, cried out, while Father Gotzens rushed about trying to calm their fears. When bombs ex- ploded nearby and the underground shelter quivered, the Catholic priest led the injured and the sick in a joint recitation of the Lord's Prayer.

76

Thousands of refugees, aware that a battle was about to engulf Cologne, were streaming westward out of the city toward American lines. Military government officers in VII Corps divisions were deluged by hordes of fleeing civilians, who clogged the roads and crowded into the little towns that dotted the Cologne Plain. Many of the disheveled refugees were bitter toward their Nazi masters. They had been told that hundreds of trucks would arrive in Cologne to carry civilians eastward, away from the advancing Americans. "But the trucks never came," said an elderly woman who had fled the small town of Kerpen. "So we headed toward the Americans on foot."

In Kerpen, one fur-coated female refugee with three leashed Scotties, which she had somehow managed to bring from Cologne, was wailing because the Nazis had taken her two automobiles. A 55-year-old, distinguished-looking man wearing a neatly tailored but threadbare suit looked around aimlessly as the jostling throngs of refugees streamed into Kerpen. He had been a respected justice of a high court in the Rhineland until the previous fall, at which point his post had been summarily abolished, and he was told that there would no longer be court appeals in the Third Reich.

Many of the refugees were sick, some with typhus. Most were hungry. All were dazed, uncertain, and gravely worried about their fate. There were numerous young women in their eighth or ninth months of pregnancy; others were carrying infants and holding children by the hands.

Major Lester A. Ahroon of Bloomington, Illinois, a military government official with the 8th Infantry Division, needed help in controlling, housing, and feeding the civilians. About 1,000 of them were milling about the little town. Ahroon sent word for the Kerpen chief of police to report to him at once. The chief showed up decked out in a full Nazi party uniform, complete with decorations and swastikas. Ahroon was aghast. "Get that goddamned silly-looking costume off before one of our men shoots you!" the American major bellowed. The policeman blanched and hurried off to change into civilian clothes.

From hundreds of windows in Kerpen, women and children peered out as the mechanized might of American armor poured through the town, headed for battle closer to the front. White sheets dangled from many windows, mute acknowledgment of the crushing defeat that was being inflicted upon the Fatherland.

It was nearing midnight on February 27 when General Terry Allen, the forwardmost leader of the 104th Infantry Division, and a few aides picked their way through the darkness to the CP of the 414th Infantry Regiment at Sindorf. Elements of Allen's Timberwolves, along with other contingents in Collins's VII Corps, were to assault the Erft Canal complex in a few hours. It had been relatively quiet at this forward command post until 2:40 A.M., when Timberwolf artillery began pounding German positions on the far banks. General Allen would direct the canal assault from this forward point.

Ahead of the 414th Infantry CP, Lieutenant Colonel Joseph M. Cummins, Jr., leader of the 2nd Battalion, which would lead the crossing, was conferring with Colonel George A. Smith, assistant division commander, and Colonel Anthony J. "Tony" Touart, leader of the 414th Infantry. Those in the battalion CP heard the heavy rush of air that foretold a large artillery shell's arrival. All instinctively ducked. Moments later there was an enormous explosion. The battalion command post was destroyed and Colonels Smith, Touart, and Cummins were killed.

Within minutes, General Allen appointed new commanders for the 414th Infantry Regiment and its 2nd Battalion, and at 3:00 A.M., precisely at H hour, the Timberwolves charged across the water barrier. Some went over in assault boats, others splashed through shallow areas on foot, and a few even swam across.

Elsewhere along the Erft, Major General Bryant E. Moore's 8th Infantry, Rose's 3rd Armored, and Major General Walter E. Lauer's 99th Infantry Divisions pushed over the canals. Resistance was light at most places, coming mainly from mortars and roving SPs (self-propelled guns). Only at Mödrath was the defense determined. Led by a resolute German major, a hodge-podge force held elements of the 8th Division at bay for two days. The Luftwaffe struck the Erft crossing sites repeatedly after dawn, going after bridges being put in by Collins's engineers. But it was a feeble effort. The Germans seldom

uld throw more than a single warplane at a time into an attack. By
ndown on March 1, Collins had 40 bridges stretching across the
ft.

s the moment had neared for the assault over the Erft complex,
alter Richardson and Sam Hogan, the lieutenant colonels and
endly rivals who led task forces of the 3rd Armored Division, were
coming increasingly thirsty—not for the case of Scotch the
vinner" over the canal had been promised by General Rose, but for
e honor of beating the other man to the far side. At jump-off, Task
rce Hogan rolled over the Erft at Glesch, using a partially de-
royed footbridge for its armored infantry. At Paffendorf, Task
rce Richardson waded through water and scrambled over twisted
idge wreckage to reach the east bank. So close was the time of the
ossing by each task force that the Richardson-Hogan "race" was
lled a tie. General Rose would have to come up with two cases of
:otch, one for each contestant.[2]

Another 3rd Armored Division task force, this one led by Lieuten-
at Colonel William B. Lovelady, was approaching the little town of
errendorf. Inside a Roman Catholic church, some 500 German civil-
ns had gathered to pray and await the arrival of Lovelady's men.
'aiting with the solemn-faced assemblage were 87 elderly members
the Volkssturm who had defied orders to draw weapons at the local
azi headquarters and defend the town to the last man.

As the tired old men of the Volkssturm, fear evident in their faces,
led out of the church with trembling hands upraised, a grimy, be-
hiskered 3rd Armored tanker called to a comrade, "I never thought
d feel sorry for a Kraut. I feel sorry for those old bastards!" The
:her nodded solemnly in agreement.

One family living in a comfortable house on the main street of Ber-
:ndorf declined the sanctuary of the Catholic church. A middle-aged
;erman, his somewhat portly wife, and their attractive 20-year-old
aughter chose to remain where they were as the guns of Colonel
ovelady's task force drew closer. They were an average middle-class
;erman family with a peppery little dog that was part dachshund.

Lovelady's tankers and infantrymen, swarming over Berrendorf,

79

entered the house where the German family had chosen to remai
The Americans drew back in shock. Dangling side by side from
rafter with heavy ropes knotted about their necks were fathe
mother, and daughter, purple-faced and tongues protruding. In o
corner the mongrel dachshund sprawled lifeless.

Turning away from the ghastly sight, the GIs picked up a no
neatly placed on a nearby table. It was written by the once-pret
daughter, whose name was Elizabeth, to a young man who apparent
had been her boyfriend and was now in the German army. How d
she expect this message to be delivered to him? the Americans muse

It was a curious suicide note, devoid of rancor or hysteria, wit
only a hint that the entire family would kill themselves in a few mi
utes, after first destroying their pet. The hand-written message rem
nisced: *How nice it was when you used to come and visit. I hope you will n
hate us for what we have decided to do.* A second note, written by th
father, was curt: *We cannot live in shame any longer.*

Two of the 3rd Armored men who had made the gruesome disco
ery took a final glance at the purplish, dangling cadavers and we
outside to be sick.

Duel at Adolf Hitler Bridge

On March 1, the eyes of all American commanders were on the Rhine bridges. None counted seriously on a span being left intact to seize, but there was always the possibility that an armored column might advance to the broad river barrier to Germany's heartland and find a bridge still standing.

All along the positions held by Bill Simpson's Ninth Army, the breakthrough after crossing the Roer River on February 23 had turned into a rout. Tank-infantry task forces were barreling ahead at a pace of 6 to 10 miles each day. No longer did the surrounding countryside resemble a battlefield; many towns and villages had been left unscarred. In one town the electric lights were burning when the Americans arrived; in another the streetcars were running. Still, there were numerous clashes where die-hard bands of Germans fought to the end.

By sundown on February 28, Isaac White's 2nd Armored Division and an attached regiment of Major General Robert L. Macon's 83rd Infantry Division had reached a point only seven miles from the Rhine. The next day, March 1, a regiment of Major General Charles

Gerhardt's 29th (Blue and Gray) Infantry Division stormed into München-Gladbach, the largest German city yet assaulted by Allied forces. It would prove to be one of the weirdest actions of the war for GIs accustomed to fighting for every foot of German soil.

The resistance, in this case, came from isolated detachments of nondescript Germans who fought briefly at streetcorners. Those who were left surrendered quickly—and with relief. Here and there an enemy self-propelled gun crashed around a corner, fired a quick round, turned, and fled. Gerhardt's infantrymen made their way from house to house while German families sat quietly in their bunk-furnished, candlelit air-raid cellars.

At one point, GIs instinctively kicked in the door of a large bank. They need not have gone to the trouble. The door was unlocked and the bank open for business. The Americans were greeted by a bank officer who spoke fluent English and had visited the United States twice. He was obviously a cultivated citizen, now deathly afraid of the grimy American fighting men prancing around his bank.

"What do you think of Adolf Hitler now?" an American lieutenant asked in a tone devoid of rancor or sarcasm.

This patently intelligent man in his mid-50s shook his head sadly and replied, "Hitler is a much misunderstood man."

The lieutenant shot back, "Yes, I for one, misunderstood the bastard!"

"Ah, yes, you are ironic," responded the neatly groomed banker. "But you will see, history will bear me out, he is one of the world's greatest leaders."

Seeing that the American officer was accepting this assessment without signs of anger, the German continued: "It is merely that Hitler has been badly advised, very badly advised, by swaggering louts with pistols on their hips, brutes and beasts, the Gestapo and others. That is how he went wrong. Now he is *kaput*."

The lieutenant shook his head in resignation, shouldered his carbine, and strode out of the bank. Outside the streets were quiet. A file of about 20 German prisoners, guarded by a lone GI, was passing, hobnail boots clattering on the hard-surfaced street.[1]

That same afternoon, General Dwight Eisenhower was jeeping into

München-Gladbach with Bill Simpson. Suddenly, a roar erupted as scores of antiaircraft guns began firing furiously at a high-flying ME-262 jet fighter. The supreme commander gazed skyward in fascination—it was the first jet he had seen. Within seconds hundreds of jagged steel fragments from exploding shells began thudding into the ground around Eisenhower and Simpson. These sharp metal chunks from sky bursts could split open a man's head—and sometimes did. For one of the few times in the war, Eisenhower put on a steel helmet.

The supreme commander and Omar Bradley stayed overnight on March 1 at Simpson's CP in Maastricht, Holland. Texas Bill and his Ninth Army staff were in high spirits. They were convinced that their spearheads could seize a bridge over the Rhine, or, denied that, Ninth Army could leap the historic river barrier on the run, using its own assault boats and bridging equipment. Simpson proposed crossing between Mundelheim and Düsseldorf, then dashing for the key city of Hamm—an outlet for the shipment of products to the German armed forces—located along the northern rim of the highly industrialized Ruhr Basin.

Simpson and his staff were convinced that, catching the Wehrmacht by surprise, Ninth Army assault troops would be opposed on the far shore only by scattered, disorganized, and rag-tag formations. Once Ninth Army was entrenched on the east bank, the Rhine would be eliminated as a formidable obstacle to the heartland of the Third Reich. All this was pointed out to Eisenhower.

The supreme commander showed intense interest in Simpson's proposal, which the tall Texan interpreted as a green light from SHAEF to bolt across the Rhine in an impromptu assault. Hardly had Eisenhower departed to the cheers of soldiers calling out "Hi, Ike!" than Simpson had bundled up his plans for crossing the Rhine and hurried to Field Marshal Montgomery to seek his approval.

But it was not to be. Montgomery turned him down flat. Simpson was stunned. "A crossing near Düsseldorf would involve Ninth Army in the industrial wilderness of the Ruhr," the 21st Army Group commander told Simpson.[2]

At Ninth Army headquarters, Simpson's staff was bitter over

Montgomery's curt rejection of their Rhine crossing plan. They railed far into the night at the British commander. All were convinced that Montgomery had rejected the proposed bolt over the water obstacle in order to keep the spotlight on the extravaganza Montgomery had long been preparing for—crossing the Rhine on a broad front in a massive set-piece action—which was being billed as the crowning glory of the field marshal's professional career.

That morning of the first day of March, Simpson had given orders to the 2nd Armored and 83rd Infantry divisions to jump off and head hell-bent for the Rhine. Their objectives: four bridges, three at Neuss and a highway bridge a short distance downstream at Oberkassel. A task force fought its way into Neuss that night, but found that all three bridges had been destroyed.

Another tank-infantry team sought to seize the Oberkassel Bridge through a deception. Iron battlewagons of the 736th Tank Battalion and the 643rd Tank Destroyer Battalion were camouflaged to resemble German tanks. Riflemen of the 330th Infantry of the 83rd Division scrambled onto the disguised tanks, with a German-speaking soldier seated at the front of each one to respond to any enemy challenge. Other infantrymen would trail on foot under orders to be as "inconspicuous as possible." In the darkness and mass German confusion, it was hoped that the task force could slip through to the Oberkassel Bridge.

Moving steadily forward, the task force had gone several miles when the shadowy figures of armed men were discerned approaching along the road from the opposite direction. The Americans tensed, but kept going. The German column—about a company—reached the head of the disguised American force and plodded on past in the blackness only 20 feet away on the other side of the road. No one in the American contingent or German column spoke or broke stride.

Dawn was starting to break as the task force neared the town of Oberkassel without having been detected. Not a shot had been fired during the long jaunt. Minutes later a file of German soldiers was spotted marching out of Oberkassel and in the direction of the advancing American force. Suddenly, a soldier riding a bicycle pointed a finger and excitedly shouted: "*Amis! Amis!* [Americans!]"

Pandemonium erupted. Their true identity revealed, American tankers and infantrymen opened a withering fire, and the startled German grenadiers began dashing wildly about in search of cover. Above the din of the heavy gunfire, the Oberkassel air-raid sirens began to moan. With all weapons firing, the American column lurched ahead and dashed for the big bridge over the Rhine. As the first tanks came into view of the span, an enormous explosion rocked Oberkassel—the Germans had blown the bridge.

North of Neuss at Krefeld-Ürdingen another Rhine bridge was intact, and at 8:00 A.M. on March 2 the entire 84th Infantry Division, led by the 335th Infantry, jumped off from a point eight miles from Krefeld with the intention of veering around the north of the city and heading for the Rhine span. The structure had a particular lure for the Americans due to its name—the Adolf Hitler Bridge. Paced by tanks rolling at full throttle, the column plunged into St. Tonis, two-thirds of the way to Krefeld, almost without opposition.

White sheets hung out of almost every window in St. Tonis. For the lead battalion, the next six hours in the town were something of a holiday. One company took over a beer hall, and while a perspiring barkeep dashed about the room serving the GIs, a buxom blonde *Fräulein* played the piano and sang.

At 6:00 P.M. the respite was over. The Railsplitters shouldered weapons and packs and, paced by tanks of the 771st Tank Battalion, marched out of St. Tonis, bound for the Rhine. Just outside Krefeld, an industrialized city with a peacetime population of 160,000, the carefree advance ground to a screeching halt. Withering bursts of fire raked the foot soldiers and tankers. Enemy grenadiers were holed up in reinforced cellars being used as pillboxes, and machine guns were spitting from three concrete air-raid shelters.

Tanks were called forward. Crews of three German anti-tank guns allowed the tanks to move past their positions, then opened up on the armored column from the rear, knocking out four tanks and a jeep. The entire attack was stalled.

It was nearing midnight when Railsplitter infantry and tanks smashed into the German positions, and this time the column plunged into a suburb of Krefeld. The town was quiet. It was taken so quickly and unexpectedly that the municipal water and electric systems were still working.

Lieutenant William B. Wood was walking down a Krefeld street at the head of the infantry company he commanded when he spotted a camouflaged German staff car approaching from the opposite direction. "Open fire!" Wood shouted. Several of his men peppered the vehicle with rifles and Tommy guns, and the staff car screeched to a halt. With Wood's men gathered around the vehicle and weapons leveled in its direction, the door slowly opened and out stepped Colonel Siegfried von Bruske, hands raised.

The German officer told Lieutenant Wood that he was a rear echelon officer and that he had never been in combat. He had been ordered to go to the front, take command of a *Kampfgruppe* (battle group), and defend Krefeld. He located Krefeld, but never did find his combat command and doubted now if one even existed any longer.

A Railsplitter battalion had been ordered to take a roundabout route to Krefeld. The truck-borne foot soldiers and 771st Battalion tanks were fired on by a few Germans with burp guns at the outskirts of Kempen, but charged on into the town as the enemy grenadiers fled. The American column roared through the deserted streets to the square in the heart of Kempen where Sunday Mass was in progress. Hearing the commotion of the tanks and trucks, the parishioners came out, knelt on the church steps, and alternately waved white handkerchiefs at the passing Americans and prayed loudly.

On the afternoon of March 2, various headquarters in Bolling's 84th Infantry Division were in a state of euphoria. With Krefeld in American hands it was only three miles to the Adolf Hitler Bridge at Ürdingen. Aerial reconnaissance reported that the lengthy span was intact and that the Wehrmacht was still using it. Alex Bolling and his commanders were aware that the bridge would probably be blown as the Railsplitters approached. But that made no difference. Plans had already been drawn up for an infantry-tank task force to bolt toward the Adolf Hitler Bridge and either cross it or, if it was blown, to ferry over the Rhine and seize a bridgehead. If needed, aircraft would drop food and ammunition on the far bank.

The dash for the Rhine bridge would begin at 2:00 P.M. The tankers and foot soldiers knew that they were launching an historic event—the first crossing of the wide, swift-flowing river by a hostile force since 1806. The tanks and trucks loaded with infantry were lined up and ready. General Bolling was on hand at the head of the column to see his men off. A sergeant in the lead tank leaned out of his hatch, thrust a cigar at Bolling, and said, "Here, General, smoke this! You can give me one across the Rhine!" Bolling took the cigar, grinned, and waved.[3]

Minutes later, General Bolling received an urgent telephone call. He was heartbroken by the message: a corps boundary line had been changed, and the ball would be given to Isaac White's 2nd Armored Division and two battalions of the 95th Division's 379th Infantry Regiment to launch a broken-field dash to the Adolf Hitler Bridge. Bolling's Railsplitters would continue their advance in a zone farther north and reach the Rhine at Homberg. Instead of three miles to the river, the 84th Division would have more than 11 miles to go.

Aware of the looming threat to the bridge named for Adolf Hitler, the German commander in the sector, General of Paratroops Alfred Schlemm, leader of the First Parachute Army, rushed the remnants of the 2nd Parachute Division—three understrength battalions—to defend the river span. Schlemm's tough young paratroopers arrived at Ürdingen only one step ahead of Combat Command B of the 2nd Armored Division. Hoping to keep the Wehrmacht from blowing up the bridge, Colonel Sydney Hinds, leader of the 2nd Armored force, ordered the 92nd Armored Field Artillery Battalion to take the structure under harassing fire. Beginning at nightfall, using shells with proximity fuses which exploded before striking the span, the antipersonnel shelling would continue for 15 hours.

Colonel Hinds for several days had been determined to seize a bridge over the Rhine. He was aware of Field Marshal Montgomery's strict order forbidding a crossing at this point and time, but if possible he intended to bolt on over a bridge—and to hell with higher authority. Once across the 1,640-foot Adolf Hitler Bridge, Hinds doubted that the British commander, in the glare of a certain worldwide spotlight, would order the American force to abandon its east-bank bridgehead. And, the colonel knew, at Ninth Army headquarters he would be hailed as a hero.

Now, with Combat Command B arrayed before the Hitler Bridge, Hinds called his commanders together and outlined his plan of action. An infantry company would dash across the span and overwhelm the Germans on the far side. While this attack was in progress, Captain George Youngblood of the 17th Armored Engineer Battalion would lead a group of his men onto the structure to locate and disarm demolition charges. Colonel Hinds was aware that his chances for success were minimal.

Expecting to encounter mainly rag-tag troops, and unaware that Schlemm's paratroopers had moved into position in front of Adolf Hitler Bridge, elements of Hinds's command jumped off at 2:00 A.M. on March 3. The spearheading tanks immediately ran into a buzz saw of counterfire, and four of them were quickly knocked out, blocking passage for other tanks in the column.

Early in the afternoon, Hinds's tanks again attacked and, approaching from the south, reached the vicinity of the bridge. There the German paratroopers pounded them with mortar fire, raked supporting infantry with automatic weapons, and destroyed several tanks with anti-tank guns and panzerfausts. The American assault again ground to a halt. Tantalizingly close to the bridge, Hinds's infantrymen could not push the remaining few yards to dash over the structure as planned.

Captain Youngblood, the engineering officer, decided not to wait for the infantry. By night, Youngblood and five of his men slipped through German positions and edged onto the bridge. The engineers moved stealthily across the span, cutting every wire they could locate, expecting to be blown up at any moment. Perspiring profusely from exertion and tension, the Americans worked their way to the far bank undetected.

"Okay," Youngblood whispered, "let's get the hell on back." They retraced their steps over the bridge, again slipped through positions of the German paratroopers, and headed for Colonel Hinds's CP. There Youngblood reported that the bridge was intact and that he and his men had snipped every wire they could locate.

Shortly after dawn, the American tankers and infantrymen were ready to rush the bridge once more. Before they could jump off, a tremendous explosion shook the terrain. The Adolf Hitler Bridge lifted into the air, then settled down into the swift waters of the Rhine. Either Youngblood and his men had overlooked the critical wires or the Germans had gone back onto the bridge in the hours before daylight and rewired the dynamite charges.

On the afternoon of March 4, with elements of Simpson's Ninth Army perched along the Rhine at several points, Major General Raymond S. McLain, commander of XIX Corps, was standing on the bank of the river a short distance north of Düsseldorf, a major industrial city on the east bank. Peering through binoculars across the 400-yard-wide Rhine, McLain could see no sign of enemy activity. The road into the heart of Germany apparently was wide open.

McLain, an Oklahoma City banker in civilian life, had risen through the ranks in peacetime as a National Guard officer with the 45th Infantry Division. He had gone overseas late in 1942 as a brigadier general and assistant commander of that division, and fought with it in North Africa, Sicily, and Italy. During the 1944 battles in France, McLain had been promoted and given command of the faltering 90th Infantry Division. He had performed so well in that post that he had been elevated once again, to command of the XIX Corps.

Now General McLain hurried from the banks of the Rhine and placed an urgent phone call to his superior, General Simpson, telling him of the beckoning opportunity to cross the Rhine north of Düsseldorf. Again Simpson rushed to Field Marshal Montgomery, telling him of McLain's call and the fact that there did not appear to be any German defenses across the river in that locale. Simpson started to explain the situation by pointing to a map he had brought along, but Montgomery declined to look at it. For the second time in three days, the British 21st Army Group commander rejected Simpson's plea for permission to bolt over the Rhine, saying he wanted to stick to his strategic plan for a broad-front crossing three weeks in the future. Montgomery indicated that the final crushing of the German armies in the west could better be achieved by the methodical sledge-hammer blow he had long ago conceived, rather than for individual elements to bolt over the river barrier and run the risk of being chopped up piecemeal.

Simpson's proposal to leap the Rhine added new fuel to the fires of acrimony between high-level American and British commanders. Montgomery's staff was convinced that the Americans were deliberately seeking to antagonize their chief by making it appear that Montgomery lacked boldness. Most suspected Patton as being the behind-the-scenes instigator of the latest Allied dispute.

For their part, Omar Bradley, Courtney Hodges, and Patton, as well as other American generals, were furious. Patton was especially irate, ranting that this was yet another effort by his antagonist Montgomery to "hog the spotlight."[4]

Meanwhile, with one Rhine bridge after the other crashing into the water from German demolitions, the dreams of American commanders to secure a usable span over the historic water barrier were fast evaporating.

At sundown on March 1, General "Lightning Joe" Collins's VIII Corps of Hodges's First Army was two miles beyond the Erft Canal complex and astride the main highways leading from Düren and Jülich on the Roer River to Cologne. That night Collins issued the orders for the final dash to the river's edge. Maurice Rose's 3rd Armored Division was to make the main effort, breaking out of the Erft bridgehead and racing for Worringen, eight miles north of Cologne on the Rhine. The Queen City itself would be captured later, and Major General Walter E. Lauer's 99th Infantry Division was to mop up bypassed pockets behind Rose's tanks. The 3rd Armored would be crossing generally flat terrain, pastoral, dotted with villages and small towns—ideal tank country.

On the right of Collins's corps, Allen's 104th Infantry Division and Moore's 8th Infantry Division were to fight their way through terrain broken by many lignite ("brown coal") surface mines with steep, clifflike sides. Abandoned mines had filled with water, creating large lakes with only narrow passages between them.

Prior to daybreak on March 2, the 3rd Armored jumped off—and promptly ran into fierce resistance. A hodgepodge of German units and stragglers battled from behind anti-tank ditches, concrete obstacles, road blocks, and dug-in machine-gun positions. Rose's tankers registered gains by the end of the day, but they had not broken through German defenses as planned.

That afternoon, Major Haynes W. Dugan of Shreveport, Louisiana, assistant G-2 (intelligence officer) of the 3rd Armored, was riding forward in a half-track not far behind the leading tanks. At the wheel was Corporal Lawrence J. Shuit of Midlothian, Illinois, and between the two men was a dog of indeterminate ancestry. Shuit had adopted the pet many weeks before, and it had been a constant companion since then.

Along the side of the road, Major Dugan spotted a curious-looking object. It was a human foot, bare and immaculate. There was no body, not even a boot. Another minor but horrific mystery in the fog of war, Dugan reflected at the time.

The major and Shuit pulled into the little town of Niederaussen hard on the heels of Task Force Richardson (Lieutenant Colonel Walter Richardson). They drove directly to General Rose's forward CP, located in a small hotel on the northeast corner of the town's main intersection. A church with a tall steeple—ideal for artillery observers—and a graveyard were across the intersection on the southeast corner.

Climbing out of his jeep, Major Dugan glanced around. His experienced eye didn't like what it saw. "If Kraut artillery ever had an ideal target to zero in on, this CP is it," he mused to himself. Main intersection. Church steeple. Both excellent aiming points.

Gunfire erupted just ahead as Dugan strolled into the CP and saw Lieutenant Colonel Wesley A. Sweat of Jacksonville, Florida, pinning a large map to the wall and circling the objective for the day with a crayon. Sweat, 3rd Armored operations officer; Lieutenant Colonel Andrew Barr of Urbana, Illinois, Division G-2; Dugan; and a few others were studying the map when the old hotel was rocked by German shells exploding outside. Those in the room instinctively dropped to the floor as shrapnel clattered against the old hotel.

A short time later, when the shelling died down, the CP men regained their feet. They stared at the wall map in disbelief: a large shell fragment had flown in through a window and imbedded itself in the map—directly in the crayoned circle marking the 3rd Armored's objective for the day.

"Do you think the Krauts are trying to tell us they know where we're heading?" an officer remarked with a straight face.[5]

All day on March 3 General Collins had been receiving reports from fighter pilots that Germans were fleeing over the Rhine at Cologne in ferries, small craft, and anything else that would float. At dusk, Collins revised his battle plan. He told General Rose, whose forward elements were four miles from the Rhine, to continue to the river at Worringen at dawn but at the same time to divert a task force southeast and "slam on into Cologne."[6]

Rose had no intention of waiting until daylight to jump off. He promptly contacted Colonel Prentice E. Yeomans, leader of the 83rd Armored Reconnaissance Battalion, and told him: "Yeomans, crank up and get on the road. Take out for the Rhine—immediately."

Minutes later, Yeomans's force shoved off into the darkness. Infantrymen led the way, followed by light tanks, armored cars, and tank destroyers. The route took the task force over flat fields and between battered little towns. At one point the quiet night air was pierced by the screeching sound of a battery of *Nebelwerfer* (multi-barreled mortars, called Screaming Meemies by the Americans) being fired to their right. Yeomans's foot soldiers flopped to the ground, but the salvo of German rockets fell well short.

The column stole onward. At Roggensdorf the reconnaissance soldiers bumped into a deeply entrenched German company and by-passed this enemy strongpoint by veering northward. On reaching

Hackhausen, Yeomans's men captured 300 surprised Germans and a battery of wheeled artillery pieces still hooked up to prime movers.

It was shortly after midnight in Hackhausen when Colonel Yeomans ordered a four-man patrol to probe ahead and to "keep going until you run into something." Those selected for the mission were Lieutenant Charles E. Coates, Lieutenant Lawrence E. Grey, Staff Sergeant Paul H. Julian, and Sergeant Aaron Muller. They hurriedly blackened their faces and set off.

A cold wind from the north was howling, causing the men to shiver constantly. The night was relatively dark. Edging ahead cautiously through a field, Lieutenant Coates in the lead spotted the dim silhouettes of a column of armed men heading in the patrol's direction. The Americans got noiselessly to the ground, lay stone-still, and held their breaths as the file of German soldiers trudged past at a distance of only 10 yards.

Coates and his men scrambled to their feet and pressed forward. They reached a road which they knew from previous map study paralleled the Rhine. Here they would try to steal over to the other side. The hush of the night was disturbed by the throbbing sound of an oncoming heavy vehicle, and the four Americans leaped back behind a clump of bushes by the side of the road. They could detect the shadowy outlines of German soldiers in the back as a truck rolled past. Somewhere out in the darkness, not far away, could be heard the chatter of German voices. Then a putt-putt-putt sound wafted through the night air as an enemy motorcyclist chugged by the Americans.

"Okay, let's go," Lieutenant Coates said in a hoarse whisper. Silently pulling themselves to their feet and listening momentarily, the men slipped over the road. A short distance later the patrol reached a bluff. Coates peeked at the luminous dial of his watch. It was 4:20 A.M. Stretching out majestically below in the dark gray of a gathering dawn was a wide, swiftly flowing river. Courtney Hodges's First Army banner had been planted on the edge of the Rhine.

9

Bloody Road to the Wesel Bridges

Baron Waldemar von Oppenheim, descendant of a venerable German family, was a wealthy Cologne banker whose spectacular, 220-year-old ancestral home, Schloss Schlenderhan, stood a short distance east of the Erft Canal complex. Herr Oppenheim's bank had continued to flourish over the years even though he was viewed with antagonism by Nazi officialdom for what it considered a serious flaw—his father had been half Jewish.

Schloss Schlenderhan contained 40 rooms, 10 baths, and had a stable of racing horses that was said to be one of Europe's best. Even with the Third Reich crumbling on all sides in March 1945, the stable still was home to 40 mares, 8 studs, and 22 yearlings who were pampered and exercised each day by a crew of grooms and stableboys.

Von Oppenheim had been a force in German politics until Hitler and the Nazis came to power in 1933. When Hitler had launched his anti-Semitic program, von Oppenheim decided to change the name of

the family's old financial institution in Cologne from Sal. Oppenheim
Companie to one with a more "Aryan-sounding" title.

Late in 1944, Waldemar von Oppenheim was forced to go into hiding after learning that his arrest by the Nazis was imminent. Himmler's SS, which had long had its eye on the tree-lined, finely manicured acres of Schloss Schlenderhan, promptly took over the castle, appropriating some of the finest racing horses in the stable. Just prior to the surprise German offensive in the Ardennes the previous December, Field Marshal Walther Model, who was in command of that action, had held several staff conferences in Schloss Schlenderhan. The headquarters of a German panzer division had been located in the sprawling building but had been pulled out hurriedly after the Americans charged over the Roer River in February, leaving behind a small force to defend the building.

Now, at 4:00 A.M. on March 2, Captain Walter Leigon of Clifton, Texas, was leading his F Company of the 104th Infantry Division up a gentle slope toward Schloss Schlenderhan, its classic contours outlined in the darkness. Neither Leigon nor any of his weary infantrymen had ever heard of Waldemar von Oppenheim; Schloss Schlenderhan was merely another objective to be cleared of its German defenders.

As the Timberwolf company was stealing toward the building, the night quiet was shattered by bursts of automatic-weapons fire from the gatekeeper's house, where a few Germans had been posted. Leigon's men concentrated a fusillade of bullets at the house and the German weapons fell silent. The Americans pushed forward and were greeted by heavy firing from the right wing of the castle and the trenches to its front.

Leigon held up his men and called for artillery fire on Schloss Schlenderhan, and in minutes the dark sky glowed as salvo after salvo of shells exploded on and around the large structure. Tanks then arrived at the scene, and in the gray of a gathering dawn Leigon and his foot soldiers rushed the castle.

Private First Class Robert D. Green of Center, Texas, was standing at a window at one end of the stables behind the castle when he spotted a German captain, clutching a rifle in one hand and a potato-masher grenade in the other, slipping around the corner of the stable. Waiting until the German was directly in front of him, Green aimed and pulled the trigger of his Garand semi-automatic rifle. Expecting a loud bark from the weapon, he heard only a click. The rifle had jammed. Green tossed the weapon to one side, rushed the enemy officer, and threw him to the ground with a flying tackle. Two other Feldgrau rushed to the aid of their captain. As Green wrestled on the ground with the German officer, he called to two of his own comrades, who dashed to the scene of the scuffle. Sergeant Manuel Garcia of Los Angeles, California, and another GI grabbed the two enlisted Germans and pinned them to the ground. The three enemy soldiers, disarmed and hauled to their feet, promptly produced surrender pamphlets, signed by General Eisenhower, which had been dropped over German positions; they claimed the documents entitled them to adequate food and fair treatment. A search of the three German soldiers revealed that each had a batch of the surrender pamphlets in his pockets for use as toilet paper.

Leigon's F Company dashed into the 40-room mansion, combed it room by room, and flushed out 58 Germans. The Americans also fell heir to 300-plus civilians, including members of the household staff and their families, together with relatives from the nearby village of Quadrath who had sought shelter in Schloss Schlenderhan's cellars. On the grounds were sprawled the bodies of 25 Germans killed during the battle, and the carcasses of several expensive race horses hit by mortar and artillery shells.

For the next few hours, the grimy, bewhiskered infantrymen of F Company took over Baron von Oppenheim's ornate mansion and, for a short time, lived the life they had always dreamed of. Slumped in overstuffed chairs before a fireplace with a roaring blaze, the GIs listened as Private First Class Thomas Boles of Greenville, Tennessee, played classical music on a huge Steinway. Boles had dreamed of being a concert pianist and had wound up as an infantryman in Germany. Suddenly, Boles broke out with a popular tune of the day, "Deep in the Heart of Texas," and the ceiling shook with the voices of baritones and tenors—not good but loud. The massive vocalizing con-

nued as budding concert pianist Boles switched to "Dance with the
Dolly with a Hole in Her Stocking."

With both Rose's 3rd Armored and Allen's 104th Infantry Divisions
driving for Cologne on March 4, Lieutenant General Friedrich Köchling's LXXI Corps, defending the sprawling city, was pulling back in
disarray. Köchling's corps consisted of only two worn-out, decimated
divisions, the 9th Panzer and 363rd Infantry. The previous day the
German commander knew that Cologne was doomed. He had stood
on a low ridge six miles northwest of the large city and through binoculars watched remnants of his 9th Panzer Division being steam-
rollered by the 3rd Armored Division.

General Köchling was forced to flee his advance command post as
Rose's tankers came into view. Köchling pulled back all the way into
gutted Cologne, setting up a CP a half mile north of the Hohenzollern
Bridge. Nearby were the towers of the cathedral. Dismayed that the
troops expected to defend Cologne itself were nowhere to be seen,
Köchling took over command of the beleaguered Queen City. He received word from higher headquarters that "One thousand Volks-
turm, armed with panzerfausts, are on the way to Cologne." Sixty
old men and young boys, with only a handful of rocket-launchers
among them, showed up.

Meanwhile on March 4, Field Marshal Montgomery's troops had
cornered some 50,000 German soldiers in a narrow bridgehead opposite the Rhine town of Wesel. These enemy soldiers, commanded
by General Schlemm—a tough man if there ever was one—were the
last remaining Germans west of the Rhine north of Cologne.
Schlemm's bridgehead was 16 miles long and only a few miles deep.
It extended from the vicinity of Xanten to Orsoy on the relatively
high ground in this generally flat terrain. His was a hodgepodge of
troops, many of them nevertheless first-rate, who had battled the British and Canadians so tenaciously in Operation Veritable for the previous four weeks. Weary and decimated, Schlemm's contingents
included remnants of four paratrooper divisions and the once elite
Panzer Lehr and 116th Panzer divisions. Only in the north against

General Henry Crerar's Canadians was there a solid bridgehead line. With their backs to the Rhine River, Schlemm's 50,000 men had only two escape paths—rail and road bridges leading to Wesel on the east bank.

Due to the sparser resistance faced by Major General John B. Anderson's XVI Corps of Simpson's Ninth Army, a decision was made for that corps to go after the two bridges at Wesel. The plan was to drive northeastward and seize Rheinberg, then turn north and try to capture the two intact river spans.

General Baade's 35th Infantry Division and an attached combat command of the 8th Armored Division jumped off for the Wesel bridges on March 4. Halfway to Rheinberg, Baade's men came under heavy automatic weapons, small-arms, and mortar fire which slowed the advance to a crawl. Seven tanks of the 784th Tank Battalion poked into a small village, and three were quickly knocked out by enemy grenadiers firing panzerfausts. An American rifle company cleared the village and rescued the four surviving tanks, but that night a German force slipped back into the little town, cornered a 35th Division platoon in a hotel, and tossed hand grenades through the windows before melting away into the darkness.

Another of Baade's battalions came under heavy fire at the base of a hill. Men of the leading company scrambled for cover in ditches and houses. The company artillery observer had lost his radio, so there was no contact with his guns. A pair of German tanks rumbled down the road toward the Americans, and a blast from their guns killed the company commander, Captain Daniel Filburn, and a platoon leader, Lieutenant John H. Hartments.

The two enemy tanks systematically shot up the houses and ditches in which the American company had taken shelter. Thrown into confusion by the loss of two leaders and the unopposed tank-shell fire, the besieged infantrymen fled. Later they were reorganized and, with tank destroyer support, moved back into and held the position.

In the same sector, a 35th Division patrol came under intense fire from a hill. Borrowing six half-tracks and several tanks from the 8th Armored Division, the battalion commander, Major Harry F. Parker, mounted one of his companies on them and sent the little task force harging into houses at the foot of the enemy-held hill. Parker's force ized the hill and another one nearby, capturing 200 Germans.

Despite the heavy opposition encountered by his division all along he line of advance, General Baade was still hopeful for a sudden reakthrough to the Rhine. Baade told Colonel Edward A. Kimball, ader of the 8th Armored Division's Combat Command B, to jump ff, seize the small town of Rheinberg, only two miles from the ighty water barrier, and to "keep rolling, push on across the Rhine nd establish a bridgehead on the other side." It was a tall order with he Germans resisting so savagely.[1]

Shortly after dawn on March 5, Kimball's task force moved out. Destination: the Wesel bridges, 13 miles away. Spirits were high. By ightfall Kimball's men would make history by being the first hostile orce since Napoleon's legions in 1806 to cross Germany's sacred Rhine River. According to American intelligence reports, only three elf-propelled guns and fewer than 300 demoralized enemy troops, nost awaiting a chance to surrender, stood between Kimball's combat ommand and the Wesel bridges.

An infantry force under Lieutenant Colonel Morgan G. Rose-orough pushed steadily ahead against virtually no opposition, cleared he town of Linfort, and by noon was on the outskirts of Rheinberg. The road to the Wesel bridges appeared to be wide open. Minutes ater, however, Colonel Roseborough received an alarming report rom the captain leading a reconnaissance company: "All hell has bro-en loose outside Rheinberg!"

Colonel Kimball contacted Major John H. Van Houten, one of his ask force leaders, and hurriedly outlined the situation in front of Rheinberg. If a dash were to be made for the bridges at Wesel, the tubborn German resistance at Rheinberg would have to be quickly wiped out. "Pass through the pinned-down recon men, attack and eize the town," Kimball ordered.

Major Van Houten split his task force into three columns and di-ected them to jump off immediately, fighting their way into Rhein-erg from the north, west, and south. Van Houten himself would oin in the assault and direct the battle from the turret of a Sherman ank.

The leader of Van Houten's force trying to pry open Rheinber from the south, Captain David B. Kelly, radioed for infantry hel after his column bumped into a large contingent of dug-in infantr and lost two tanks to panzerfaust rockets. But before infantry suppor arrived, Kelly decided to risk a quick rush into town with tanks alone Three Shermans charged forward; German anti-tank guns knocke out all three.

Kelly decided to try again with his remaining tanks; he would lea the attack in his Sherman. The other tanks lagged, and Kelly race into the narrow, winding streets alone. The isolated Sherman, wit throttle wide open, dashed into the center of Rheinberg and, wit treads squealing on the cobblestone street, circled a World War statue of a German soldier in the town square as its machine-gunne cut down a soldier taking aim at Kelly's tank with a panzerfaust.

By now the German defenders had been alerted that a lone Amer ican Sherman was dashing through the streets of Rheinberg, an Kelly's tank had to run a gauntlet of fire as it raced back out of town Five times anti-tank gun shells barely missed Kelly's steel-plated vehi cle, and as he neared the edge of town Kelly and his crew twice wer jolted as flat-trajectory projectiles crashed into the Sherman but faile to halt it.

Returning to the few tanks remaining in his force south of town Captain Kelly found that two companies of infantry had arrived Only about an hour of daylight remained, but Kelly was determine to take the town. With Kelly himself leading the attack on foot, th infantry and the handful of surviving tanks headed for Rheinberg.

Heavy fighting broke out almost immediately. Kelly's tank-infantry force struggled forward and knocked out three 88-millimeter guns an five 20-millimeter antiaircraft weapons, and captured over 100 pris oners. Darkness had fallen, so Captain Kelly's men dug in for the night on the southern outskirts of Rheinberg.

As Kelly had jumped off to assault Rheinberg from the south, task force leader Van Houten had sent a column of light tanks under com mand of his executive officer, Major Edward Gurney, to enter the town from the west. Gurney quickly ran into impossible resistance

nd put in an urgent call to Colonel Kimball: "Nine tanks knocked
ut. In danger of being overrun unless help sent at once."

Colonel Kimball realized the situation was desperate. Major
Gurney was a resolute fighting man and would not make an urgent
ppeal for reinforcements unless his force was in danger of being
viped out. Kimball, frantically scraping together as many in-
antrymen as he could locate, hustled them into half-tracks. Kimball
imself jumped into the first half-track, and with a grinding of gears
he rescue contingent lurched forward.

A short distance ahead the motorized force reached a blown bridge.
Colonel Kimball leaped from his half-track, shouted at his men to
lismount, and started forward on foot. Nearing Major Gurney's
talled column, Kimball and his scratch force were raked by machine-
un and rifle fire, and pounded by mortars. Through a pall of smoke
overing over the battleground, Kimball saw a sight off in the dis-
ance that sickened him—the nine blackened hulks of Gurney's
knocked-out tanks, the charred bodies of crew members hanging gro-
esquely out of hatches.

Colonel Kimball kept marching until he reached Gurney, who was
preparing another charge into Rheinberg with his remaining 18 tanks
nd three half-tracks. Kimball hopped into one of the half-tracks and
ordered his little force of infantrymen to join in the attack. The tank-
nfantry team had begun moving toward Rheinberg when suddenly
he column was raked by Germans in camouflaged pillboxes on either
side of the road. Colonel Kimball leaped out of his exposed seat in the
half-track and scrambled into a light tank. "Get the hell going!" Kimball shouted at the driver. "Catch up with those other tanks ahead!"

After traveling only a few hundred yards, the combat command
leader and the light tank crew received a terrific jolt, and the tracked
vehicle halted in place. It had been struck a glancing but disabling
blow by an 88-millimeter shell. The vehicle started to burn as Kimball and the others bailed out of the hatch and, with machine-gun
bullets looking for them, scrambled into the roadside ditch.

The ditch was already filled with Gurney's men, many of them
wounded. Major Gurney himself was among them, his tanker's suit

saturated with blood from a gaping wound in his stomach. Along th
road several of Gurney's tanks and half-tracks were burning fiercely
The din was ear-splitting—the incessant fire of German machin
guns, the violent explosion of mortar shells, the concussions of pan
zerfaust rockets. Wounded Americans in the ditch were screaming fo
medics. "We'll all be killed!" a despairing voice shouted above th
turmoil.

Colonel Kimball spotted a house about 50 yards away and, pullin
himself up into a crouch, vaulted out of the ditch and started racin
for the structure. Hard on his heels was one of his men. Germans i
the pillboxes had spotted the two running Americans and went afte
them with 88-millimeter and machine-gun fire. Kimball and his com
panion scrambled through a cellar window and, gasping for breath
stretched out on the floor.

"Well, we made it," the soldier observed with a trace of surprise.

"Yeah, we made it," Colonel Kimball responded. [2]

Kimball knew that his Combat Command B of the 8th Armored
Division had been badly chewed up. He had seen the carnage first
hand. But as yet he was unaware of the disaster the Germans had
inflicted on CCB—it had been almost wiped out at Rheinberg
Thirty-nine of Kimball's 54 tanks had been knocked out, 92 of his
men had been killed and wounded, and 31 were missing.

Paratrooper General Alfred Schlemm's men, with the waters of the
Rhine only two miles behind them, had slammed shut the open gate
on the road to the key bridges at Wesel.

Meanwhile, General Bolling and his men of the 84th Infantry Division had put aside their dejection over having the opportunity to be
the first across the Rhine jerked out from under them at the last minute by a change of corps boundaries. Now elements of the Railsplitter
division were charging toward two other spans in their new sector to
the north—the Baerl Bridge and the 2,050-foot-long Admiral Scheer
Bridge connecting Homberg on the west bank with Duisburg.

Dawn was breaking on March 4 as a battalion of Railsplitter infantry jumped off from positions two miles west of the Rhine; by noon
they had reached the village of Meerbeck. There the leading elements

ould see the battalion objective, the Baerl railroad bridge. Peering through binoculars, officers saw a steady stream of German military traffic pouring east across the Rhine. Artillery fire was called for and n minutes sky-bursts were exploding over the bridge, halting all but occasional groups of foot soldiers risking a dash to the far shore.

Approaches to the Baerl Bridge were heavily defended by automatic weapons and anti-tank guns. After dark, all three companies in the Railsplitter battalion fought their way forward and reached the Rhine at 8:00 P.M. At 8:07 there was an enormous explosion and a brilliant flash of light spread over the river and landscape. The Baerl span had been blown.

At 10:00 that night, March 4, Captain Oreste V. Valsangiacomo was at the head of his company spearheading another 84th Division battalion's thrust to seize the Admiral Scheer Bridge at Homberg. It was again very dark. Valsangiacomo halted his marching men and went into a roadside house to check his map. For some reason, he chose to enter through the back door. Inside, he heard German voices. The captain froze and listened. The chatter was coming from an adjoining room. Valsangiacomo silently stole over the floor for several feet until he could gain a glimpse of the room. A surge of alarm spread through his being. Six German officers were seated around a table and by the dull glow of an oil lamp were poring over maps.

Captain Valsangiacomo realized instantly that he had stumbled into a German command post, which meant that his entire company had unknowingly infiltrated well behind enemy lines. The fact that the German CP had not posted guards around the house indicated that the enemy officers inside thought they were a safe distance from American troops. The captain stealthily made his way back out the rear door, dreading that with each step the Germans in the CP might be alerted to his presence and send a fusillade of bullets into his back.

Returning to his men lounging along the road outside, Valsangiacomo quietly issued orders to continue the advance. As he was obviously in the midst of German positions, the captain felt his chances were better to move forward than to try and march back through enemy lines. The company moved off into the blackness, leaving the Wehrmacht officers in the CP deliberating on how to stop the American push to the Rhine in that sector.

Minutes later the stillness exploded as enemy machine-gun and rifle fire began to pepper the intruders. The Americans scattered for cover and in the confusion became separated. At dawn, Captain Val sangiacomo had only 30 men with him. He quickly rounded up the remainder of his company and, along with two other companies of his battalion which had pushed forward in support, headed the short distance to the Rhine. They reached the shore at 9:45 A.M. and saw their objective, the massive Admiral Scheer Bridge—a twisted wreckage half-submerged in the water. The Germans had blown the structure during the night.

Thirty miles south of Wesel on the previous night of March 3, Staff Sergeant George Dove of the 414th Infantry Regiment of Allen's 104th Division had slogged through a forest west of Cologne with his L Company comrades. As he stumbled along, burdened by the infantryman's customary heavy gear, Dove reflected that it was the densest, blackest forest he had ever seen. It was cold and damp— typical "dogface weather." Sergeant Dove's company had been fighting hard all day and now was moving ahead on the heels of the fleeing Wehrmacht soldiers.

Suddenly, L Company emerged from the blackness of the woods and stumbled into the outskirts of Buschbell. Soon the weary Americans arrived at a huge hothouse. Dove stared at it in fascination. "What a crazy, mixed-up world," he thought. Outside, raw wintry weather; inside, flaming red peonies blooming. "If only I could send a bouquet of those beautiful peonies to Mom," flashed through the young sergeant's mind. "Or better still, let her know that I'm still alive."[3]

Dove and other fighting men of the Timberwolf Division were battling their way through elaborately constructed defenses ringing Cologne. There were four semi-circles of trenches and strongpoints in and around the besieged city. Throughout the cold, gray day of March 4, intermittent snow flurries and icy rain made life miserable for soldiers on both sides. Terry Allen's men pushed ahead against poradic resistance and by nightfall were within two miles of Cologne.

North of the Queen City that same day, Lieutenant Colonel Bill Lovelady's task force of the 3rd Armored Division attacked and captured the Rhine town of Worringen, the spearhead's original primary objective in the assault over the Roer River. Ancient Cologne, one of Germany's largest rubble piles, now was in its death throes. On the morrow, VII Corps commander Lightning Joe Collins planned to send Rose's 3rd Armored into Cologne from the north, Allen's Timberwolves driving into the city from the west, and Moore's 8th Infantry Division pushing eastward to reach the Rhine south of Cologne. The coup de grace would be launched at dawn.

The great river city, which Adolf Hitler had decreed should be held to the last man and the final brick, was devastated. Most of the heavy vehicles, guns, and tanks that had fuel had rumbled over the massive Hohenzollern Bridge to the east bank. Left behind were what the Feldgrau called "suicide squads." Many of these reluctant warriors had no intention of being listed in Reich newspapers in the black-bordered columns headed, "Fallen for the Führer," but a number of them had vowed to fight to the death.

Ferry boats hauling troops, equipment, and vehicles to the far side of the Rhine had been pounded by American artillery for two days. A gray-haired German priest, his black cassock muddy and threadbare, had been walking near Hohenzollern Bridge the previous day. The lengthy span was crowded with more than 200 civilians, pitiful refugees fleeing to the east side. As the priest froze in horror, the center section buckled, swayed, and plunged into the Rhine, sending the crowd of screaming civilians to their deaths.

That night German army demolition teams blew up the remainder of the bridge. The last chance for Cologne civilians to escape the doomed city had vanished.

As night drew its ominous cloak over Cologne on March 4, a pall of smoke drifted lazily over the city. Standing guard over the debris were the twin spires of the cathedral, spared on the orders of General Joe Collins. The 515-foot Gothic spires, Collins had decreed, would not be used to register artillery.[4]

10

"Mighty Cologne Has Fallen"

At dawn on March 5, Cologne lay mute, gray, and ghostlike. The crescent-shaped Queen City of the Rhine, once famed for its culture, the beauty of its magnificent churches, its spacious boulevards, its sprawling industries and immense railroad yards, was hemmed in on three sides by troops of Lightning Joe Collins's corps. The wake for the departed city was already underway as Collins's fighting men set about administering the last rites. Great billows of smoke were spiraling into the slate-colored sky on both sides of the cathedral from a fuel dump that was burning out of control. There was no one to fight the blaze.

There was a great stir of activity outside the city as the 3rd Armored and 104th Infantry divisions got ready to assault Cologne from five directions, through the suburbs of Junkersdorf, Mungersdorf, Binkendorf, Longerich, and Ossendorf. The American fighting men would encounter seemingly endless piles of rubble. Since Britain's Royal Air Force heavies had made their first 1,000-plane strike on Cologne in May 1942—the first instance of saturation bombing—Cologne had been shaken under 42,000 tons of bombs.

North of the once-bustling metropolis, Colonel Leander L. Chubby" Doan's Task Force X of the 3rd Armored Division was on the road shortly after daylight, heading south. Doan, whose nickname referred ironically to his thin, angular frame, was a Texas rancher who had been renowned in peacetime as a capable polo player. His men felt it was poetic justice that Doan be selected to spearhead the drive into Cologne. His Task Force X had been out in front in the 3rd Armored's drive from Normandy, across France, and into Belgium and Germany. Colonel Doan himself had become a "public enemy" in the Nazi scheme of things and had a price on his head for "war crimes"—meaning he and his tankers and armored infantry had killed a lot of German soldiers in combat.

Task Force X, a mixture of Sherman medium and light tanks, half-tracks, armored cars, jeeps, and infantry in trucks, poked into the northwest suburb of Binkendorf at 7:10 A.M. after brushing aside sporadic opposition and an occasional sniper. At an airport, tanks of a task force led by Lieutenant Colonel Bill Lovelady encountered 16 dual-purpose 88-millimeter guns, and after a spirited fight the formidable weapons were overrun. Yet another task force commanded by Colonel John C. Welborn ran into heavy fighting from "suicide squads" while seizing several towns on the outskirts before pushing on into Cologne itself.

Lieutenant Hugh U. McBirney, leading a platoon of 3rd Armored Sherman tanks, was engaged in duels with concealed anti-tank guns and snipers burrowed into the rubble piles. One of his tank commanders, Sergeant John Burleson, spotted a sniper and ordered his gunner, Corporal Hubert Foster, to fire at the German with the tank's 76-millimeter gun.

"Did'ja get him?" Burleson called out.

"Goddamned if I know," Foster replied, "but he was standing behind that wall—and now there ain't no wall."[1]

Rose's tankers and foot soldiers drove steadily forward, scrambling over wasteland that had once been a beautiful park or an ornate apartment complex or block of palatial homes. At one point the Germans had constructed a roadblock of five streetcars, reinforced by steel rails driven into the street. The Shermans buttoned up, charged the strong point, and barreled through it. White sheets hung from many win-dows. Dazed civilians, ignoring sniper and occasional mortar fire, had crawled out of their cellars and were standing along rubble-strewn streets to watch the Americans pass by. Other civilians chose to flee the dying city. Refugees, pushing carts, baby buggies, and small wag-ons piled high with blankets and other household goods, were strag-gling forlornly along the streets and roads, clogging the paths of 3rd Armored's advance. The terrified civilians gazed in despair as shells sent up tall plumes of white smoke in the distance.

Most of the refugees were old people, children, and the infirm, and here and there a mother with two or three small children. One young woman had a baby in one arm, a loaf of bread in the other. Children were carrying plates and cups, crying as they stumbled along in the icy wind and driving rain. They often screamed as rifle fire erupted.

Maury Rose's tankers and foot soldiers fought on toward the center of Cologne.

Earlier that morning, at about the time the 3rd Armored Division was jumping off to plunge into Cologne from the north, B Company of the 87th Mortar Battalion was furiously firing its 4.2-inch mortars just outside the city limits on the west. The mortarmen were shooting WP (white phosphorous) shells in order to lay down a smoke screen to shield the Timberwolves of L Company of Colonel Gerald C. Kel-leher's 415th Infantry Regiment. The L Company men would have to cross two miles of flatland, devoid of cover, to reach the first houses of Cologne.

There was subdued excitement among the men of the 4.2-inch mor-tar company. The unit had made the assault landing on D-Day in Normandy, fought in countless battles across the face of Europe, suf-fered heavy casualties, and was now within sight of the famed Dom (cathedral) on the Rhine. Most B Company veterans had doubted they would see this day. Teenaged Private First Class Alphie Gregoire of Biddeford, Maine, observed to a comrade, "If my lucky rabbit's foot doesn't fail me for a while longer, I'll make it to the Rhine." He felt his shirt pocket to see if the good-luck charm was in place.[2]

It was approaching 8:00 A.M. when Lieutenant Mervin Tieg, an infantry platoon leader in the Timberwolves' L Company, called out

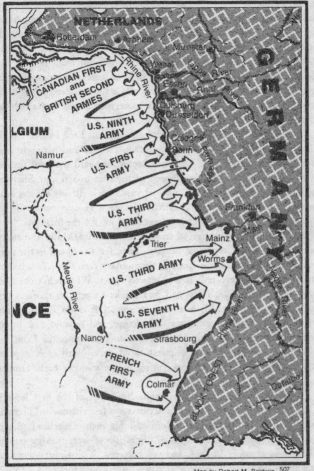

Destroying Germans West of the Rhine

Map by Robert M. Baldwin
507
504

to his men, "Okay, let's go!" Tieg's platoon would lead the assault into Cologne. Leaving the concealment offered by a cluster of houses in the village of Welden, Tieg and his men deployed into a skirmish line and set out across the flatland. They were especially tense. Ahead lay 4,000 yards of open field before the Timberwolves could reach the first houses in Cologne. A lot could happen in two miles. The thin early morning haze and the smoke screen to their front helped. But the smoke could not halt a German shell or deflect machine-gun bullets. Besides, the smoke was hovering only a few hundred yards to the front. Tieg and his men would have more than a mile and a half left to go after they advanced through the smoke.[3]

Out in front was a squad led by Staff Sergeant Fred Hoover. If the Germans opened up through the smoke, Hoover and his eight men would be cut down first. The squad sergeant heard the loud revving of motors and instinctively glanced to the rear to see four Sherman tanks rumbling along a few hundred yards back. It was always reassuring to have friendly tanks nearby, Sergeant Hoover reflected, but this assault over the flatland was basically a job for the infantry.

Hoover and his men moved steadily along. It was chilly and rainy, but the men were perspiring heavily. They felt naked out in the open field. Now the squad was through the smoke screen the 4.2-inch mortar company had created. It was ghostly quiet. Why didn't the Germans start shooting? They had had weeks to zero in mortars and artillery. Hoover and his men kept moving. Off in the distance they could see the first gray houses of Cologne, and beyond that in the drizzle and haze loomed the majestic, 515-foot towers of the Dom.

The staggered line advanced at a rapid pace. With each step another frightening thought began to haunt the mind of each Timberwolf: would he set off a Schu mine?[4]

Hoover and his squad were within 200 yards of the first houses. The sergeant thought briefly of the carnage a few entrenched German machine guns would make of him and his men. Overhead, flimsy American Piper Cubs, the eyes of the artillery, were making passes. There was comfort in that, Hoover reflected. The Krauts wouldn't dare fire with those "snoopers" overhead. Or would they?

One hundred yards from the first houses, Sergeant Hoover called out, "Okay, let's rush 'em!" Burdened with combat gear, the infantrymen began running forward and scrambled into the first buildings without a shot being fired at them. Most of the houses were empty and locked. The Timberwolves battered in the doors or shot off the locks; they were solidly established in the suburb of Junkersdorf.[5]

At about noon that day, March 5, a despairing General Friedrich Köchling, who had recently been appointed to defend Cologne, was seated in his CP in the cellar of a battered building near the Hohenzollern Bridge. His orders had been: Defend Cologne to the last. With what? Köchling reflected. There was no organized resistance. Scattered throughout the pulverized metropolis were bands of grenadiers and individual panzers, who would fight to the end or surrender promptly, depending on the makeup of the individual soldier.

Heavy footsteps were heard on the steps leading down to Köchling's CP. Two SS officers, stern-faced and formal, advised Köchling that he was under arrest for dereliction of duty and treason. He was to return to Berlin immediately to stand trial on the charges. In the chaotic climate of a crumbling Nazi Germany, General Köchling could have simply surrendered himself to the Americans now swarming about Cologne. Instead he placed himself in the custody of his chief of staff, crossed the Rhine, and headed for Berlin and almost certain execution. In Adolf Hitler's courts Köchling's long record of battlefield courage and accomplishments would be ignored. He had "permitted" the American juggernaut to crash into Cologne.[6]

At daybreak on March 6, Timberwolves of the 414th and 415th Infantry regiments began their steady, grim advance through the rubble-strewn streets, bound for the banks of the Rhine. This morning the advance was more violent than usual. Terry Allen's men had to be extra cautious, as mortar, rifle, machine pistol, and self-propelled guns sporadically started up against the Americans.

Private First Class William C. Buck of F Company, 415th Infantry, and his comrades reached Cologne's famed Opera House late in the afternoon. The ornate old structure had been heavily damaged, but

Buck and several others found suitable, dry quarters in the cellar. To the astonishment of all, the lights came on when a curious soldier flipped a switch. "Goddamned!" called out a GI, "I'd about forgotten what an electric light looked like." Investigation revealed that the building had its own power plant, which had survived the heavy bombings.

Outside the Opera House, solemn GIs were erecting a wooden sign with hand-painted letters in German and English on a large rubble pile:

> GIVE ME FIVE YEARS AND YOU
> WILL NOT RECOGNIZE GERMANY
> —Adolf Hitler

The Führer had made that promise to the nation while campaigning for Germany's chancellorship in the early 1930s.

At 3:00 P.M. that day, C Company of General Rose's 36th Armored Infantry Regiment was advancing street by street through the gutted city. Its objective: the cathedral. Lieutenant Robert J. Cook of Wellsville, New York, C Company commander, doubted if his task force of tanks and infantry could reach the church by darkness. Harold Siegman, photographer for an American publication, was following close behind Cook's forward elements. As with other cameramen, his eye was on the twin towers, an internationally recognized landmark. Without the Dom, Siegman had no picture. So Siegman headed back to the First Army press camp, planning to return at dawn.

In the meantime, Lieutenant Charles Rhodes, a Signal Corps photographer, and three of his cameramen decided to keep tagging along behind Cook's task force. So did another civilian photographer, Fred Ramage. The little bevy of cameramen had their instruments ready just in case there was a break in the heavy fighting on the approaches to the cathedral. At 5:00 P.M., in the gathering gloom, the German opposition suddenly melted, and Lieutenant Cook's tanks charged ahead, one Sherman halting directly in front of the cathedral.

As though in deference to the fabled religious structure before which they had halted, Lieutenant Cook's tankers shut off motors. An eerie silence settled over the Dom Platz. The commander of the tank directly before the Dom eased open the hatch cautiously; snipers might well be around. This was precisely the type of Cologne picture the cameramen had been waiting to snap for days. No need for a caption: the first American tanker, silhouetted against the mighty cathedral, easing out of his Sherman turret to survey the once bustling metropolis laid waste around it.

The cameramen leaped out of jeeps, dashed beside the Sherman, adjusted lens settings, and raised their instruments. The 3rd Armored tanker, aware of the historical significance of the photographs about to be snapped, stepped out of the turret and onto the top of the tank. He removed his tanker's helmet, wiped his grimy face with his sleeve, and surveyed the devastation.

The eager American photographers snapped their shutters at the precise moment when a concealed German tank fired its gun. The panzer crew had been biding its time, waiting for the right moment to shoot at point-blank range. The results were the most dramatic and horrible pictures of the war—the tanker's leg, severed just above the knee by the shell, was captured on film flying through the air.

The tanker screamed in pain and toppled onto the cobbled street. As if on cue, all other tank turrets in the Platz slammed shut and the battle resumed. The legless soldier was struggling on one knee and the bloody stump. His shrieks of agony could be heard above the roar of tank gunfire that had erupted in front of the cathedral.

The photographers, caught up in their work, ignored the shot and shell flying around and kept their cameras clicking and rolling. Two of those suddenly plunged into the center of the fierce tank battle were Sergeant Leon Rosenman and Corporal James Bates, both 3rd Armored photographers who were shooting motion pictures. They kept cameras grinding, first sighting on the menacing Panther tank down the street, then switching toward the Shermans and one of Lieutenant William L. Stillman's new Pershings in the square.

The Pershing commanded by Sergeant Robert Early had a higher-velocity gun than that on the Sherman. Early's Pershing quickly exchanged armor-piercing rounds with the Panther, and suddenly the

American tank scored a direct hit. The Panther burst into a flaming torch in the shadow of the cathedral—the motion-picture and still photographers recording the entire savage scene.

Meanwhile, medics scrambled and slithered over the cobblestones to the tanker whose leg had been shot off and applied a tourniquet to the stump. Litter bearers snaked toward the soldier, who continued to shriek. The screams became fainter and finally there was a gasp, then silence. By now the German Panther had been knocked out, and the litter carriers trotted away with the dying 3rd Armored tanker.

Just before the tank battle erupted in the Dom Platz, Major Haynes Dugan, assistant G-2 of the 3rd Armored Division, and several war reporters had been searching for a vantage point from which to see if a bridge over the Rhine had been blown. The group scrambled up into the tall steeple of a Catholic church a few blocks from the cathedral just in time to see the last span over the Rhine erupt in smoke and plunge into the river. But they got an unexpected bonus—a bird's-eye view of the savage duel between Pershing and Panther.

Silence returned to the Dom Platz.

Surrounded by a sea of devastation, the cathedral still stood—defiant or benign—the only significant edifice in the entire sprawling city to have escaped the wrath of American and Royal Air Force bombs and the arrival of "Lightning Joe" Collins's artillery. On seven occasions during the 100-plus bombing raids on Cologne over the years, projectiles had damaged the ancient stonework. The floor of the massive structure was littered with broken masonry, but the cathedral was still basically intact and all of its religious treasures were safe, having been covered with solid brick shields or removed to the interior of Germany.

John H. "Beaver" Thompson, the veteran correspondent of the *Chicago Tribune* who had been covering the war since parachuting into Tunisia with the 509th Parachute Infantry Battalion during the invasion of North Africa in November 1942, was starting to walk out of the cathedral when a little group approached. Two American riflemen, Corporal Eugene J. Binek and Corporal John Milas, both of Chicago, were herding three male civilians.

"Anybody know where they're rounding up civilians now?" Binek called out. Thompson did not know. Milas looked around the desolate cathedral interior and tilted his head back to scan the dim ceiling. "Jesus, this is a big joint," observed the soldier. He gave a low whistle, then the two Americans resumed their trek with the civilians.

Outside, a Royal Air Force officer asked Thompson, "What do you think of the railroad station?"

The correspondent looked blank and muttered, "What railroad station?"

Pointing to an enormous pile of rubble and twisted steel, the blue-uniformed Briton replied, "Why, that one over there, old fellow."[7]

Heavy artillery shells from across the Rhine were crashing periodically throughout Cologne, which, by late afternoon of March 6, was virtually in American hands. Here and there a stubborn pocket grimly fought it out to the end, but prisoners were steadily arriving at impromptu POW cages. Lieutenant Robert G. Banbury of Aurora, Illinois, and the 104th Division, was up to his ears in the work of interrogating 75 prisoners. One young German, speaking passable English, said to Banbury, "Please, sir, can my name be mentioned tonight on BBC [radio]?"

Surprised at the curious request, the American lieutenant inquired, "And why do you want your name mentioned on a London radio station?"

"Because, sir, my father always listens to BBC."[8]

An eerie silence had fallen over Cologne, a once highly industrialized city the size of 1945 Pittsburgh or Liverpool, England. One could drive for blocks without seeing a living human. Debris filled the streets, making travel difficult. The bombs had smashed everything. Along the Venloerstrasse the familiar red facade of Woolworth's, an American chain store which the Germans had continued to operate throughout the war, was in ruins. The huge clock in front of the Cologne West railroad station was an undamaged sentinel over the plain of rubble, but its hands had stopped at quarter past eight on some undetermined date. At Willi Beucham's wine shop on Venloerstrasse, about 20 ragged civilians were looting the store.

Back in the United States that night, millions of eager Americans were tuned to the six-o'clock radio news. They were elated to hear the popular newscaster, H. V. Kaltenborn, lead off the newscast: "There's good news tonight! Mighty Cologne has fallen!"

Just west of Cologne the following day, March 7, Supreme Commander Dwight Eisenhower drove up to General Joe Collins's forward CP. Climbing out of his car, Eisenhower greeted the VII Corps commander warmly and congratulated him on the swift seizure of the Third Reich's fourth largest city—only 12 days after Collins's corps had jumped off at the Roer River. The VII Corps lightning bolt enhanced Collins's "Lightning Joe" moniker. Taking Collins aside, the supreme commander told him, "Joe, if another field army is created in the ETO [European Theater of Operations], you are going to command it."

Collins was flattered by the disclosure. At 48, he was the youngest American corps commander in the ETO, and to head an army would require his being promoted over many generals senior in rank and age. The silver-haired general told Eisenhower that while he was deeply grateful for the supreme commander's confidence in him, he would rather continue to lead VII Corps to Berlin.[9]

That day, members of Rose's 3rd Armored Division took over the *Staats Gefängnis* (state prison) in northern Cologne. The Americans were appalled by what they found in the drab facility that had housed those the Nazi government had deemed to be political enemies. There had been 800 prisoners in the *Staats Gefängnis*. Only 85 remained—all feeble, ill, and dazed. Suffering from malnutrition and beatings, racked by typhus, most of the inmates could not move under their own power. They were carried to American army hospitals.

Curious GIs from other outfits found excuses to flock into Cologne, now ghostly quiet except for the occasional crash of a German shell from across the Rhine. Signs were erected on the outskirts of the city: SIGHTSEERS KEEP OUT.

The Americans in Cologne, while defending the west bank of the Rhine in this sector, settled into a strange existence after the extreme tenseness of almost continual fighting since the Roer River attack on February 23. A platoon of Timberwolves was quartered in a mansion in what had once been a fashionable district bordering the Rhine. On the fourth floor a machine gun had been set up so that it could cover

he river and the terrain on the German side. The weapon was
manned 24 hours a day. The oil-drenched machine gun had been
placed on a once-beautiful piece of mahogany furniture, set back from
the window so that the crew could fire out the opening without being
seen. The principal sport was to try to knock a German soldier off a
bicycle at extreme machine-gun range. One Timberwolf, peering
through binoculars across the Rhine, called out, "Here comes one of
the bastards!"

Sergeant Mike Martin grabbed the field glasses. "Oh, boy. Wait
. . wait . . . now, let him have it."

Private First Class Buford Wiggins squeezed the trigger and a
stream of bullets swished across the Rhine.

"Did we get him?" someone called out.

"Hot dog!" Sergeant Martin hollered. "I don't know if we got the
son of a bitch, but that bicycle is rolling along without a rider."[10]

Another Timberwolf, Sergeant Randolph Coleman, was approaching the Club Cologne, located in the basement of a battered building
less than 100 yards from the Rhine. Coleman was greeted by the
doorman, Private First Class Turman Smith, gave the correct password, and went down the cellar steps to Club Cologne's main room.
There the host, Sergeant Eugene MacParland, was genially dispens-
ing assorted spirits and good cheer to Timberwolves at the crowded
bar. The atmosphere was almost festive.

Actually, the "doorman" was on duty with his BAR in a sandbagged
position, with a field of fire along a wide stretch of the Rhine. Genial
"host" MacParland was in charge of two machine-gun crews whose
weapons were aimed out windows in the cellar. All things considered,
Sergeant Coleman and his comrades agreed, this was the way a war
should be fought. But a mammoth 280-millimeter German shell exploded nearby, rocking Club Cologne and reminding the "patrons" that
sudden death still lurked along the river.

11

"Your Name
Will Go Down
in Glory!"

Willi Bratge, the diminutive 40-year-old captain commanding the 36-man security force at Ludendorff Bridge at Remagen, put in an urgent call on the morning of March 6 to Field Marshal Walther Model's headquarters in an effort to beef up his feeble defenses. Bratge, a wounded veteran of Poland, France, and Russia, was growing increasingly nervous. On paper at Model's command post, Bratge had 1,000 men under his control: 500 members of the Volkssturm, 220 antiaircraft troops, 180 *Hitlerjugend* (Hitler Youths), 120 undependable Russian volunteers, and the 36 convalescents of his security company, many of whom still wore bandages and required daily treatment. In reality, most of Bratge's 1,000 men had melted away in recent days.[1]

Only six members of the Volkssturm remained at the bridge site, the others having fled. A large number of men manning the flak guns on the sheer, 660-foot basalt cliff at the eastern end of the river span, known as the Erpeler Ley, had also vanished. Bratge doubted that in an emergency those who stayed behind, other than his own tiny company, would respond to his orders. In a tight situation, he could rely only on his 36 combat convalescents.

Bratge's urgent call to Model's headquarters was routed to a mere lieutenant named May, an ordnance officer at that. Remagen and its Ludendorff Bridge held low priorities in the German scheme of things, as only a military idiot, in the words of the field marshal, would attempt to cross the Rhine at that point. Lieutenant May could handle whatever was the bridge commander's complaint.

Bratge made an urgent plea for reinforcements. "The Americans are so close I can actually hear tank fire," he exclaimed.

"Nonsense!" snapped the brash young officer in Army Group B headquarters. "The Americans aren't coming to Remagen, they are bound for Bonn." Bonn was north of Remagen. Lieutenant May scoffed at the tank fire Bratge claimed he had heard. "It's just a small American force protecting the flank of the main body."

Bratge was growing angry. "I've been a soldier for a long time," he exclaimed, "and I tell you that's not a weak force I hear but a strong one!"

The bridge commander banged the telephone down in frustration.

Only the day before, on March 5, Lieutenant General ·Walther Botsch had visited Remagen to inspect defenses. Botsch had been selected by Model to straighten out the command mess in the Bonn-Remagen sector involving the bureaucrats of the *Wehrkreis*, the administrative agency, and various components of the fighting branches. Highly regarded in the Wehrmacht, Botsch seemed a natural for the job. He had a knack for getting people to work together and relished untangling insolvable situations.

One of General Botsch's first actions after his appointment March 1 was to rush to Remagen to inspect defenses. After conferring with Captain Bratge on the fifth, Botsch promised to send reinforcements in considerable strength—two battalions of grenadiers, a battalion of artillery, and a battery of 88-millimeter flak guns. Defending the bridge and the town with the hodgepodge collection available to Bratge was impossible, General Botsch had agreed.

Before departing the picturesque old Roman town, Botsch discussed at length with Captain Bratge and Captain Karl Friesenhahn, the commander of the engineer group at Remagen, Adolf Hitler's specific directions for blowing up the Rhine bridges on the approach of American and British forces. Hitler had cautioned not to demolish them so soon, as to cut off German forces trying to escape over the spans. Friesenhahn, wiry, gray-haired, nearing 50, was directly responsible for destroying Ludendorff Bridge at the precise appropriate time.

A loyal Nazi, Friesenhahn assured General Botsch that he would demolish the 1,000-foot span, but pointed to an ominous factor: he had not as yet received the explosives ordered by him many days before. Botsch assured Bratge and Friesenhahn that the explosives and reinforcements would arrive at Remagen and departed with the feeling that the situation was in good hands at the Ludendorff Bridge.

That same day west of the Rhine, the commander of the German LIII Corps was out looking for his scattered troops when he was surprised and pounced on by General Patton's tankers. Field Marshal Model immediately appointed Walther Botsch to succeed the captured commander, but not before flatly rejecting Botsch's earnest plea for reinforcements for Remagen. Troops were hard to come by, and Model had no intention of wasting them at the Ludendorff Bridge. The field marshal took it for granted that the huge double-track railroad span would be blown up in the Americans' faces.

Now, on March 6, the reinforcements promised by General Botsch still had not arrived. No one told Willi Bratge that they were not coming. Stepping outside his little office on the east bank, Bratge could again hear the tank fire to the west. The bridge security-force commander strode across the river span, where he ran into Captain Friesenhahn. Both men were in a foul mood. The engineering officer upbraided Bratge for sending his 36-man company to Viktoriaberg, a dominating hill on the west bank. Bratge was furious. He shouted at Friesenhahn that he had placed his men there to give the alarm of approaching Americans in plenty of time for Friesenhahn and his engineers to blow Ludendorff Bridge. The two glared fiercely at each other, then turned and walked away.

Friesenhahn was that rare officer who was both liked and respected y his men, although some thought he was too zealous a Nazi. He d been gassed and wounded three times in World War I and had sen to the rank of sergeant in his engineer combat battalion. After e war he had gone into private business, was drafted at the outbreak hostilities in 1939, and received a lieutenant's commission the folwing year.

Willi Bratge, 10 years younger than Friesenhahn, had been Ludenrff Bridge security commander since the previous November. Like riesenhahn, he was short, measuring five feet five inches. He had ained as a schoolteacher, but in 1924 had joined the *Reichswehr* (the 0,000-man army allowed under the World War I armistice) when bs proved scarce. Bratge was seriously wounded in Russia in Auist 1944, and upon recovering was assigned to Remagen. A profesonal soldier, he neither enthused nor complained about his rear-area b.

With American spearheads driving for the Rhine, Lieutenant Genal Otto Hitzfeld, commander of the LXVII Corps responsible for efense of the Bonn sector, called in his 32-year-old adjutant, Major lans Scheller, at close to midnight on March 6. He told the bright, ager Scheller to take command of all forces at Remagen at once. litzfeld stressed the crucial need for blowing the Ludendorff Bridge t the last possible moment. "If need be," the corps command said olemnly, "you will give the order for demolition yourself."

Scheller was delighted to be given such a significant task. He umped into his staff car and headed through the blackness for Reaagen.

he American tanks Captain Bratge had heard firing in the distance ere those of the U.S. 9th Armored Division, commanded by Major eneral John W. Leonard, which was spearheading a thrust by First rmy's General Courtney Hodges to link up with an armored divion of Patton's Third Army coming up from the south. Hopefully, he gigantic trap would snare General Gustav von Zangen's entire ifteenth Army of 245,000 men.

Leonard's 9th Armored had jumped off that morning in two parallel columns, with Combat Command A on the right and Combat Command B on the north. CCB, led by Brigadier General William M. Hoge, had plunged ahead for 10 miles that day, and by late afternoon had rolled into Meckenheim, 11 miles from Ludendorff Bridge. Hoge, a stern taskmaster, had been lashing his men forward for over a week to prevent the disorganized Germans in front of him from digging in. His troops were near exhaustion. So was Bill Hoge.

Hoge was a West Pointer, as were his two brothers and two sons. He had served in the same division with Generals Hodges and Leonard in World War I and had won the Distinguished Service Cross for bridging the Meuse River under fire. Hoge was painfully frank when dealing with officers superior in rank, a trait that had slowed his climb up the promotion ladder. The Lexington, Missouri, native never tried to win a popularity contest with his men. But they respected him for his ability, fairness, and courage.

The tall, handsome Hoge sent for his operations officer, Major Ben Cothran, in civilian life the city editor of the *Knoxville Journal*. "Ben, pick out a route to Bonn," Hoge directed. Bonn, 15 miles north of Remagen, was the birthplace of Beethoven and in recent weeks had been headquarters of Field Marshal von Rundstedt, the commander in chief, west. Combat Command A on the south was to seize Remagen and then, with the Ludendorff Bridge expected to be blown at the last moment, was to head southward.

At 6:00 P.M. General Hoge again sent for Major Cothran. "Hold up on the Bonn route," Hoge told him. "There's been a change in plans."

It was nearing 10:00 P.M. when 9th Armored Division Commander John Leonard received a phone call from his boss, Major General John Millikin, the old cavalryman who led III Corps. In a discussion of the situation in front of Leonard, Millikin observed almost casually, "See that little black strip of a bridge at Remagen? If you happen to seize that, your name will go down in glory."

After Millikin hung up, both he and Leonard promptly forgot the remark. Capturing any bridge over the Rhine, Hitler's final barrier to the heart of the Third Reich, would be one of the monumental feats of the war. Neither man thought the opportunity would evolve.

At 2:30 A.M. on March 7, General Bill Hoge was roused from fitful lumber by an aide. Colonel John "Pinky" Growdon, 9th Armored Division operations officer, had arrived at Hoge's CP with urgent new orders for Combat Command B. Hoge's exhausted troopers were to jump off at 7:00 A.M. in two columns and head for Remagen and Sinzig, a village three miles south of Ludendorff Bridge.

"What about the bridge?" Hoge inquired.

"There's no definite directives concerning it," Colonel Growdon replied. "Except you should fire on it only with time fuses [shells that exploded before striking the target]." This would disrupt German military traffic fleeing over the span but would not damage the bridge itself.

Hoge hurriedly assembled his commanders. Combat Command B would be split into two task forces. Lieutenant Colonel John Engeman, who had a reputation for being brusque, would lead his own 14th Tank Battalion and the 27th Armored Infantry Battalion, race due east to Remagen, and capture the town. An apparently more difficult mission was given to Hoge's other task force. Under Lieutenant Colonel William R. Prince, the 52nd Armored Infantry Battalion was to strike out for Sinzig, force a crossing over the Ahr River, which flowed into the Rhine, and seize the town.

In a damp cellar in a battered house in Meckenheim, one of Engeman's infantry company commanders, Lieutenant Karl Timmerman, was awakened before dawn. "The Old Man's called a meeting for oh-six hundred [6:00 A.M.]," the messenger told the drowsy Timmerman of West Point, Nebraska. Timmerman felt good—even excited. He had just had five hours of sleep, the most at one stretch in a week, and had been commander of Able Company of the 27th Armored Infantry Battalion for only a few hours. The lanky Timmerman had replaced an officer seriously wounded in the attack on Meckenheim.

Timmerman's mother back in Nebraska was a German war bride of World War I, and her three brothers were fighting with the Wehrmacht in the current global conflict. Only the day before, Lieutenant Timmerman had written his wife La Vera in Nebraska: "There's no glory in war. Maybe those who have never been in battle find that certain glory and glamour that doesn't exist. Perhaps they get it from the movies or the comic strips."

Promptly at 6:00 A.M., with his company commanders gathered, Lieutenant Colonel Engeman snapped, "Okay, break it up. Let's get down to business."

He said that he had just returned from a conference with General Hoge and that "the Old Man wants us to push ten miles an hour today." Turning to Lieutenant Timmerman, Engeman, a University of Minnesota graduate, stated, "Timmerman, you've got an important job to do. Put your doughs on half-tracks, and I'll give you a platoon of tanks. You'll be the advance guard for our whole goddamned task force."

Timmerman, all business as usual, nodded. "When do we start?"

"Move out at oh-seven hundred."

Colonel Engeman looked toward Lieutenant John Grimball, whose southern drawl had been the butt of countless jokes. "You take your platoon of Pershings [a new wide-bodied tank with a powerful 90-millimeter gun] and team up with Timmerman in the advance guard." As an afterthought: "Those Pershings scare hell out of the Krauts."

"Yeah, suh," replied Grimball, a South Carolina attorney in peacetime.

The hard-bitten Engeman dismissed his commanders with a parting admonishment: "Everybody will have to get off their asses and push hard. They tell me there's not too much enemy stuff ahead of us."

It was miserable weather when General Bill Hoge's two task forces jumped off—misty, chilly, and gray. An hour later General Leonard strode briskly into Hoge's CP in Meckenheim. "How's it going, Bill?" the division commander inquired.

Hoge looked up from a map. "John, your intelligence tells me the Ludendorff Bridge is still standing," he observed. "Suppose I find the bridge hasn't been blown. Should I take it?"

"Hell, yes," Leonard said without hesitation. "Get the hell on across it."

Yet like everyone else, Leonard felt there was no chance of seizing an intact bridge over the Rhine.

Eleven miles to the east of where Leonard and Hoge were confering, German vehicles and troops, remnants of General von Znagan's retreating Fifteenth Army, had been pouring over Ludendorff Bridge since dawn. The Feldgrau, disheveled, dog-tired, hungry, and dispirited, were thankful for one blessing: the overcast day would ground the dreaded *Jabos* (Allied fighter-bombers).

Standing by the bridge on the west side, an anxious Willi Bratge was viewing the remnants of a once-proud German army moving past. In a foul mood, Bratge periodically shouted, "Get your asses in gear. The *Amis* are right on your heels." The exhortation was greeted only with dull stares.

Now Bratge had a new worry. He had been counting on the flak guns on top of the 660-foot-high Erpeler Ley to shoot down on the Americans as they approached Remagen. Now the bridge security-force leader learned that the antiaircraft weapons on the dominating cliff had been sent, without his knowledge, south to Koblenz to meet General George Patton's thrust for that key Rhine city.

Several hundred feet up in the overcast sky, an artillery pilot-observer was searching for possible targets on the approaches to the Rhine. It was 10:32 A.M. when Lieutenant Harold Larsen's Piper Cub knifed through a low-lying cloud mass and emerged above the Rhine River. Larsen was startled by the sight suddenly unveiled before his eyes: the massive Ludendorff Bridge. Excitement surging, the lieutenant banked his Cub and headed for home to report the dramatic development.

A half-hour later Captain Bratge was pondering his course of action in his CP at the western end of the Ludendorff Bridge. He felt as depressed as the miserable weather outside. The door opened and in walked a lanky Wehrmacht officer, his eyes red-rimmed from lack of sleep, a stubble on his face. He introduced himself as Major Hans Scheller, adjutant to General Otto Hitzfeld. Bratge felt a brief glimmer of hope—no doubt the major had come with the troops and guns promised two days previously by General von Botsch.

"Where are the reinforcements?" Bratge asked anxiously.

"I have no idea what you are talking about," Scheller replied. " was sent here to take command of Remagen and the bridge."

Bratge suspected that the newcomer was a spy and demanded to se his identification. Satisfied that the young major was who he said h was, Bratge felt relief. Now the ultimate responsibility for the feeb defenses at the Ludendorff Bridge was on someone else's back.

Major Scheller promptly asked for a rundown on the final prepar tions to blow the Ludendorff Bridge and was told that 60 separat dynamite charges had been placed at crucial points on the massiv structure. There was no time to be lost. Scheller and Bratge starte across the bridge, connecting each charge with a main cable leadin into the railroad tunnel in the Erpeler Ley at the east end of the spar

At the same time that the two German officers were connecting th charges, Colonel Engeman's task force was rolling hell-bent for R magen and had reached the village of Bierresdorf, only three mil from the Ludendorff Bridge. Led by Lieutenant Timmerman's i fantrymen of Able Company in half-tracks, the column continued o ward and reached a high, heavily wooded plateau. Platoon Sergea Carmine Sabia, a stocky 25-year-old from Brooklyn, grew suspiciou The woods were too quiet. He halted the column, sent several burst of fire into the thick trees, then took nine of his men and headed o foot up the road.

It was nearing 1:00 P.M. when Sergeant Sabia came over a sligh rise. He stopped suddenly in breathless astonishment. There befor his eager eyes was a magnificent view of the Rhine River, stretchin out for miles in each direction. There was the town of Remagen—an the huge Ludendorff Bridge, still standing. For a moment he was to excited to speak. Then he blurted out, "Jee-sus Kee-ryst! Look a that!"

Footsteps were heard clopping up the road from behind. Staff Ser geant Joseph DeLisio, a sturdily built man from New York City wh sported a mustache, called out, "What in hell's holding you guys up? Then he saw the view of the Rhine—and the bridge. For seconds, h too was speechless. DeLisio had been known for his courage in tigh situations, and his comrades had said that he often dared the German the Bridge at REMAGEN!

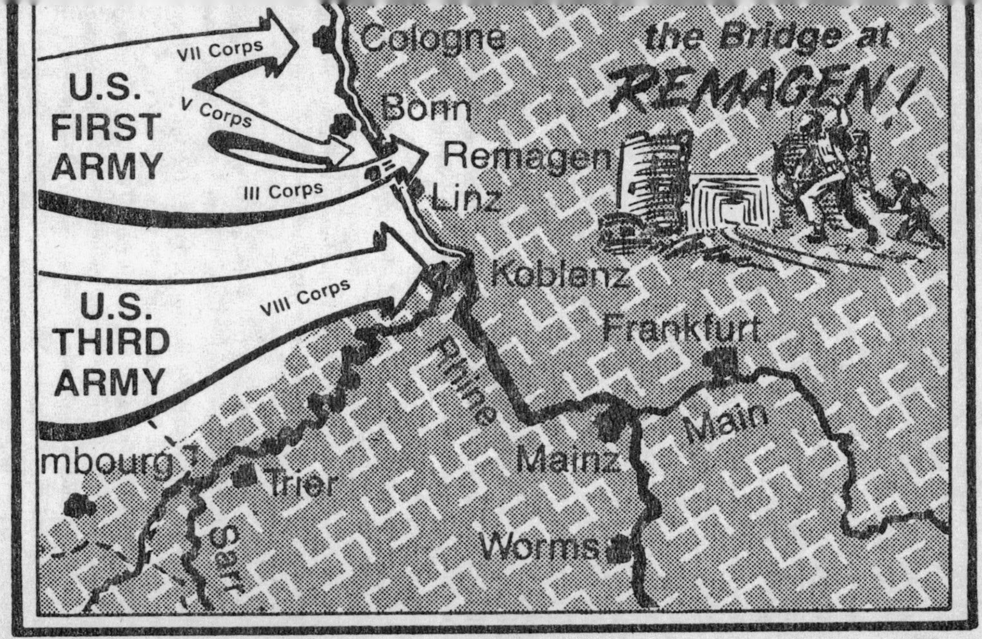

Cologne
Bonn
Remagen
Linz
Koblenz
Frankfurt
Main
Mainz
Worms
Rhine
Trier
Sarr
mbourg?

VII Corps
V Corps
III Corps
VIII Corps

U.S. FIRST ARMY

U.S. THIRD ARMY

Map by Robert M. Baldwin

507
504

The Bridge at Remagen

to try to kill him. Now, staring fixedly at the mighty Ludendo[rff]
Bridge in the valley far below, DeLisio had a sudden thought: [I]
want no part of that goddamned bridge. The Krauts will blow it [to]
hell the minute we start across."

Now Lieutenant Karl Timmerman, the Able Company co[m]-
mander, and one of his platoon leaders, Lieutenant Emmet Burrow[s,]
dashed up to the observation point. Hard on their heels the task for[ce]
commander, Colonel Engeman, jeeped forward, dismounted, a[nd]
scrambled up the road. He, too, stood with the others in silen[ce.]
Peering into the distance through binoculars, Engeman saw that [the]
bridge was clogged with German traffic and troops. He ordered [his]
artillery observer to open fire on the steel span with time fuses.

Meanwhile General Hoge's Task Force Prince had raced southe[ast]
almost unopposed, dashed over the Ahr River, and stormed Sinz[ig,]
taking German defenders in concrete bunkers by total surprise. So[me]
300 Feldgrau were hurriedly taken prisoner. A local resident inform[ed]
Lieutenant Fred De Rango that the Ludendorff Bridge was to [be]
blown at 4:00 P.M.—only about two hours away. Unable to rea[ch]
higher headquarters by radio to tell of the Sinzig native's stateme[nt,]
De Rango headed his platoon toward the bridge with the intention [of]
ripping out the charges—hopefully well before 4:00 P.M.

On the hill overlooking the Ludendorff Bridge, the incisive, stock[y,]
fast-talking Colonel Engeman began snapping out orders. Lieutena[nt]
Timmerman was to take his Able Company on foot down the win[d]-
ing road and into Remagen. C Company would follow shortly afte[r-]
ward in half-tracks.

Turning to the slow-talking lawyer from South Carolina, Lieute[n-]
ant Grimball, Engeman stated, "John, I want you to barrel dow[n]
through Remagen with your Pershings. Pound the bridge with ta[nk]
fire. Get rid of any goddamned Kraut who tries to blow it up."

It was 1:50 P.M. when two of Lieutenant Timmerman's platoo[ns]
marched off briskly down the narrow, winding road toward Remag[en,]
with Emmet Burrows's platoon taking the point. The other A Com[-]
pany platoon, with the stocky Sergeant Joe DeLisio in charge, took [a]
short cut, scrambling down a steep path past a maze of vineyards. D[e]
Lisio and his men reached the hard-surfaced Bonn-Remagen highwa[y]

which ran along the west bank of the Rhine, then turned right and headed for the town. They were greeted with a few bursts of small-arms fire, but had little trouble scampering into the first few houses.

Minutes later one of DeLisio's men dashed up to him, breathless and excited. "Hey, Joe," he called out, "we just captured a god-damned Kraut *general*."

DeLisio followed the private to a house where a squad of his men were standing about staring at a uniformed German seated docilely in a chair. DeLisio broke out laughing. "Let the son of a bitch go," he said. "You captured a goddamned railroad conductor."

As Timmerman's foot soldiers and Grimball's tankers edged through Remagen toward Ludendorff Bridge, Captain Friesenhahn and four engineers were hunched down around a demolition charge on the west side of the river span, shielded from American view by the Becher furniture factory. The dynamite was supposed to blast a hole in the road large enough to halt American vehicles and tanks on the western approach to the bridge. Friesenhahn, the devoted Nazi and dedicated soldier, waited as long as he could. A German artillery unit was scheduled to arrive at any moment to pull back over the bridge.

Now small-arms fire began hissing overhead and Grimball's Pershings were exploding shells all around. Still Friesenhahn hesitated. His nervous, perspiring men cast anxious glances at the engineer captain. The retreating German artillery unit was nowhere in sight. Friesenhahn spotted armed figures coming up the street wearing American helmets, and the menacing outlines of the low-slung Pershings. He could wait no longer. "Fire the charge!" he shouted.

An engineer jerked the cord and the five Germans scrambled for cover. Seconds later there was a mighty explosion, and a thick cloud of dust and smoke hovered over the blast. When it had cleared, Friesenhahn breathed a sigh of relief—a 30-foot hole had been blasted in the road in front of the bridge. The five Germans began dashing back over the bridge for the eastern bank, but halfway across a shell from a Pershing exploded next to Friesenhahn and he was knocked unconscious. A few minutes later he regained his senses and staggered on over the bridge.

129

Meanwhile, General Bill Hoge had raced to the observation post o[n] the hill overlooking the Rhine and Remagen. He stared in amazemen[t] at the intact bridge. General Leonard's words that morning flashe[d] through his mind: *If the bridge is still intact, take it and get the hell o[n] across.*

Suddenly Hoge shouted at his task force commander, Colonel En[ge]geman: "Grab that bridge!" The general, normally soft-spoken, wa[s] highly agitated. "Take some tanks and put them on each side of th[e] bridge and fire across the river. Send your infantry on across." H[e] brushed aside Engeman's explanation that he already had infantry an[d] tanks in Remagen.

After he had calmed down, Bill Hoge mused softly to no one i[n] particular, "It'd be nice to have a bridge."

It was 3:15 P.M. when an aide handed an urgent message to th[e] general. It was from Lieutenant De Rango at Sinzig stating that th[e] mighty Ludendorff would probably be blown in 45 minutes.

"You've got to hurry!" General Hoge shouted at the unfortunat[e] Colonel Engeman. "They're blowing the bridge at sixteen hundred."

II

Over
the
River

12

"Suicide Mission" at Ludendorff Bridge

eneral Bill Hoge, a scowl on his handsome face, was peering in-
ntly through binoculars at the Ludendorff Bridge in the Rhine gorge
elow. He could not spot any action at the river span or in Remagen.
Vith the opportunity of a lifetime in his grasp, the attack on the
ridge seemed to have totally bogged down.

Hoge spun around to Major Murray Deevers of the 27th Armored
nfantry Battalion. "Deevers, get down the hill and get your men
cross," he ordered. To Lieutenant Colonel Leonard Engeman he
aid, "I want you to get to that bridge as soon as possible."

The brusque Engeman, who had been the target of Hoge's verbal
rrows for more than two hours, snapped back, "I'm doing every god-
amned thing possible to get to the bridge!" The task force com-
ander leaped into a jeep and sped into Remagen. There he radioed
hn Grimball, the Pershing tank-platoon leader. "Grimball, get to
ell to the bridge!" Engeman rasped.

"Suh, Ah'm already theah!"

Engeman told Grimball to continue to fire on the bridge to keep th[e] Germans from any kind of activity on the structure. He ordered Lieu[?]tenant Hugh Mott of the 9th Armored Engineer Battalion to me[?] him promptly behind a hotel near the bridge.

"Mott," Colonel Engeman stated, "General Hoge wants you to g[o] out on that bridge and see if it's mined or loaded with TNT and if it hold tanks. I'll give you fire support with my tanks."

The angular, imperturbable 24-year-old Mott, from Nashvill[e] Tennessee, never batted an eye, even though he had just been order[ed] to launch a "suicide mission." If he survived German fire on the e[x]posed bridge, chances were excellent that the enemy would blow th[e] structure sky-high.

Taking two men with him, Eugene Dorland, a husky masonr[y] worker from Kansas, and John Reynolds, a diminutive textile craft[s]man from North Carolina, Mott headed for the Ludendorff span.

At the head of the bridge, Major Deevers was crouched behind [a] building conferring with one of his company commanders, Lieutenar[t] Karl Timmerman.

"Do you think you can take your men over that bridge?" Deever[s] inquired.

"Well, we can try, sir," Timmerman replied in a calm tone. He to[o] knew that it would be a suicide mission. The unflappable Nebraska[n] could see German soldiers laboring feverishly at the far end of th[e] bridge. "What if it blows up in my face?" he asked Deevers.

Major Deevers did not reply. He avoided Timmerman's eyes turned around, and walked away.

As Deevers was giving the order to Timmerman to cross the bridge[,] General Hoge on the bluff overlooking Remagen was jolted by a[n] urgent message from 9th Armored Division commander Joh[n] Leonard, who was unaware of the startling developments in progres[s] at Ludendorff Bridge: "Present mission canceled. Push south with al[l] possible speed to link up with Patton's 4th Armored Division comin[g] up from Koblenz."

Bill Hoge had one of the great opportunities of the war in hi[s]

asp. His infantry had not yet started over the bridge. He could
ll call the operation off. If he seized the bridge, he would be
oclaimed a hero and his disobeying a direct order would probably
overlooked. If he failed, or the bridge blew up in Lieutenant
mmerman's face—a likely possibility—Hoge would be court-mar-
led and sent home in disgrace, branded forever as an irresponsible
lure.

Hoge paused for only seconds. There was no time to lose. "Take
e bridge!" he snapped.

On the far side of the bridge, Captain Friesenhahn, still dazed by
e shell that had knocked him cold, stumbled up to Bratge, whom he
w in the mouth of the tunnel at the base of the towering cliff. "The
mis are at the Becher factory!" he cried out.

"Blow up the bridge!" Bratge yelled back.

Friesenhahn refused. That order would have to be given by Major
cheller. And only a short time previously Scheller had reminded
riesenhahn of Adolf Hitler's decree that any man would be shot who
ew up a Rhine bridge prematurely.

Bratge started back through the dark tunnel, now crammed with
fugees and rag-tag assortments of soldiers, to Scheller's CP at the far
d of the excavation. There he told Scheller, "You've got to blow the
ridge—now!"

The young major reminded Bratge of the Führer's dire warning.

"Well, if you don't give the order to blow it, I will!" Bratge
apped heatedly.

Scheller shrugged his shoulders resignedly. "All right," he stated,
low it up."

Bratge struggled back through the masses of people in the black
nnel, located Friesenhahn, and bellowed, "Okay, blow up the
ridge!"

Friesenhan turned pale. Who was giving the order? He hesitated.
hen he knelt by the detonating device. Soldiers and civilians in the
nnel flopped to the ground in anticipation of the tremendous explo-
on which would send the bridge flying into the sky. Friesenhahn
ave the key a quick turn, which should have detonated the 60 dyna-
ite charges. Nothing happened. He repeated turning the key. Still

135

no explosion. He quickly grasped the problem: an American shell ha
cut the cable leading to the charges.

The engineer captain asked his men for volunteers to race out on
the bridge and light an emergency charge by hand. His call w
greeted by silence. Obviously, this was a suicide mission. Finally,
sergeant named Faust agreed to try. As he edged out of the tunn
and began racing for the primer cord some 75 yards away on th
bridge, Sergeant Faust was raked by American automatic-weapo
fire and blasted by Pershing tanks, but reached his destination. The
he was blanketed in a cloud of smoke from exploding shells.

Moments later the engineer sergeant's form emerged from th
smoke and he raced back to the tunnel with machine-gun bullets ni
ping at his heels every step of the way. Another failure, thought
desperate Friesenhahn on seeing Faust dashing toward him in the tur
nel. But minutes later an explosion rocked the ground, and pieces c
bridge timber shot into the air. The men in the tunnel cheered, b
their elation was short-lived. When the black smoke began to diss
pate, the Ludendorff was still standing. Friesenhahn felt sick in th
pit of his stomach. Faust's blast had only blown a hole in the bridge
surface.

On the west end of the river span, Lieutenant Timmerman and hi
men were in shell craters ready to dash across when Sergeant Faust
explosion roared. "Hold it," Timmerman shouted. "We can't get ove
now."

Moments later someone called out, "Look, the goddamned thing
still standing!"

"Okay," Timmerman shouted, waving his arm forward, "let's go!

The company commander headed for the bridge, but his men hesi
tated. Major Deevers called out to one platoon in a cheerful voice
"Come on, fellows, let's get across. I'll see you on the other side an
we'll all have a chicken dinner."

Obscene shouts greeted his remark. The major was told what h
could do with "your goddamned bridge."

With Lieutenant Timmerman repeatedly urging his men to follow
him and waving them onward, one by one the hesitant infantrymen
scrambled out of craters and edged cautiously onto the bridge. "

on't want to go, but I will!" called out Private First Class Art
Massie. One sergeant shouted to another, "C'mon. We'll just take a
ttle walk across." Bellowed another: "Hell, who wants to live for-
ver?" Timmerman was standing upright on the bridge and calling
ut, "Move it! Move it!"

Now most of his company was on the structure and starting the
,000-foot dash to the far side. Machine-gun bullets began to sing past
he charging Americans. Standing on the bank, Chaplain William T.
Gibble was calmly taking movies of the crossing with his eight-milli-
neter camera, oblivious to the bullets lacing the air around him.

Lieutenant Hugh Mott, the imperturbable engineer, and two of his
nen were furiously cutting each cable they could locate. Halfway
ver the span they located 100 pounds of dynamite tied to a beam
inder the decking. They clipped the wires and moved onward.

The dash across the bridge seemed endless, as though the men were
unning on a conveyer. Many could not help glancing down at the
wirling waters of the Rhine, 80 feet below. With their heavy packs
nd combat boots, they wondered how long it would take for the
urrent to drag them under. Machine-gun bullets skipped past them
fter ricocheting off the deck and struck sparks from steel girders.

Leading the charge over the Ludendorff with long strides was lanky
Sergeant Alex Drabnik, an assistant squad leader from Ohio. Drabnik
was churning his legs so rapidly that his helmet fell off, but he kept
unning and reached the far shore. At the time he was not aware of
his historic achievement—the first armed and hostile soldier to fight
his way to the far side of the Rhine since Napoleon's legions had
tormed across in the early part of the previous century.

Hard on his heels was Marvin Jensen, in civilian life a plasterer
rom Minnesota. Jensen shouted repeatedly in the dash over the
bridge, "Holy shit, do you think we'll make it?" Breathless and still in
danger on reaching the far bank, Sergeant Drabnik could not resist
saying to Jensen, "Holy shit . . ." He gasped for breath. "Holy shit,
we made it."

Reaching the far end of the bridge, Sergeant Joe DeLisio of the
Bronx flung open the door to a massive stone tower, one of two that
anchored the east end of the span. He came face-to-face with five

Germans huddled over a jammed machine gun. DeLisio could n
bring himself to cut them down, but fired shots into the wall ov
their heads and shouted, "*Hände hoch!* [Hands up!]"

The startled Germans quickly raised their hands. "Any comrad
upstairs?" snarled DeLisio, still alone, in his fractured German.

"*Nein*," was the reply.

"Well, we'll see about that," the American said.

He prodded the five Germans up the circular stairway with th
muzzle of his rifle. At the head of the stairs were a German privat
apparently too terrified to get off the floor, and a lieutenant who wa
drunk. The officer leaped toward what appeared to be a detonatin
device, but DeLisio fired two shots into the floor and the inebriate
officer thought better of it.

Lieutenant Timmerman, perspiring heavily, was the first America
officer to reach the far end of the bridge. He shouted to a sergean
"Take a few men and look into that tunnel, but don't get into a fight

Unknown to Timmerman, Sergeant DeLisio and several others ha
already edged into the black, quarter-mile-long tunnel under the E
peler Ley. They treaded softly, even fearfully, expecting to be hit b
machine-gun fire at any moment—or expecting the tunnel to b
blown up, burying them under tons of earth. The only sound wa
their muffled footsteps. Suddenly, voices were heard in the blacknes
ahead. DeLisio fired two rounds into the ceiling, and two German
with hands upraised edged out of the darkness.

A few hundred yards past the east entrance to the tunnel, Sergean
Sabia led his squad to the little railroad depot at Erpel, just in time t
spot a train routinely pulling in from the north, obviously unawar
that Americans were anywhere near. Sabia and his men ducked into
trench and watched. Awkwardly scrambling down from the trai
were a number of elderly soldiers armed with rifles. A neatly dressed
crisp-actioned young lieutenant was shouting orders at the pitiful fig
ures, trying to get them into some sort of formation.

Sabia waited until the Germans were in line, then he and his me
stood up with leveled rifles and shouted, "*Hände hoch!*" The Volks
grenadiers dropped their rifles and threw their arms into the air—a
did the young lieutenant.

Inside the stifling tunnel, a melodrama was unfolding. Seeing Americans on the east bank, Captain Bratge hustled back through the tunnel to Major Scheller and excitedly told him a counterattack would have to be formed. Scheller shrugged his shoulders and gave his approval. Moving back through the tunnel, Bratge started hustling up men for the counterattack, but before he could move out of the tunnel, a soldier rushed up from the rear and said that Scheller had vanished.

The mass of civilians in the tunnel now panicked, and many began shrieking, crying, and begging Bratge to surrender. Americans were at the front and back of the underground passageway and more were pouring over the bridge every minute. Bratge, though a dedicated soldier to the end, now lost all hope. He conferred with his only remaining officers, Friesenhahn and two lieutenants. They agreed the situation was hopeless. The civilians, now bearing white flags, began to stampede toward the Americans at the front of the tunnel.

Bratge, a forlorn figure, faced his few remaining soldiers and said softly, "I command you to cease fighting and destroy your arms." After five years, Willi Bratge's war was over.

Meanwhile, another company of Major Deevers's 27th Armored Infantry Battalion had worked around to the rear of the tunnel, where a lone German sentry quickly surrendered. With Americans blocking both entrances of the tunnel, 200 German soldiers and irregulars trapped inside—including captains Bratge and Friesenhahn—gave up the struggle.

Learning that his men were on the far side of the Rhine, Bill Hoge leaped into his jeep at the observation point on the bluff and sped back to his CP in Bierresdorf. A short time later General John Leonard, the 9th Armored commander, drove up. Before Leonard could get out of his vehicle, Hoge went out to meet him. "Well, we got the bridge," he told Leonard almost casually.

Leonard's face froze in a stern mask and he snapped, "What in the hell did you do that for? Now we've got the bull by the tail, and caused a lot of trouble!" Then Leonard broke out in a wide grin. He was as aware as anyone that this had been a monumental feat of arms.

"Let's push it," he said in a serious tone, "then buck it on up to corps." The division commander added, "You've disobeyed order But you were absolutely right, and I'll back you all the way."

Bill Hoge felt all along his boss and longtime friend would feel th way, but he was greatly relieved to hear him say it.

Lieutenant Colonel Leonard Engeman had nearly his entire task force including a few tanks, on the far bank before the Wehrmacht hig command learned—and then only by chance—of the disastrous tur of events at Remagen. Günther Reichhelm, Model's 31-year-old ope ations officer and reputedly the youngest colonel in the Germa army, had just reached Army Group B's new headquarters after was pulled back east of the Rhine.

Reichhelm was shocked to hear an antiaircraft officer telling of th Remagen events. He could not locate Model, so he began furiousl telephoning to find a commander near Remagen who could mount prompt counterattack and wipe out the American bridgehead.

Reichhelm reached Lieutenant General Joachim von Kortzfleisch a Bensberg, 20 miles north of Bonn. Kortzfleisch had been in charge c rear defenses along the Rhine, manned only by scattered groups c the old men of the Volkssturm. The general considered his comman a pitiful joke. Now he was being placed in charge of wiping out th Remagen bridgehead by Colonel Reichhelm.

"Wipe it out with what?" Kortzfleisch demanded to know.

"You can use the Eleventh Panzer and Panzer Lehr divisions, Model's aide stated. Both were first-rate outfits, but badly decimate and short on fuel.

Getting these large formations to Remagen would take time. Wha was needed, Kortzfleisch knew, was a fighting unit near the scene o the American penetration. The general and an aide, Colonel Rudol Schulz, were driving toward Remagen when they spotted what migh be salvation. Parked on a village street near Bonn was a full equipped battalion of infantry and 16 panzers loaded with gasolin and ammunition.

The unit's commander, Lieutenant Colonel Ewers, was eager t take on the assignment. But he told General Kortzfleisch that hi orders were to go to Bonn. Kortzfleisch tried frantically over th telephone to get Ewers's orders changed and launch the primed and waiting battle group against the Americans at Remagen. He was unsuccessful. Finally he got through to Field Marshal Model. "If Ewers and his experienced men don't throw back the Americans tonight," Kortzfleisch declared, "then it must be presumed that the inner door to Germany will remain open for the Americans."

Model was unmoved. He said that he knew all about the situation and had discussed it with the Führer. Neither Hitler nor Model considered Remagen that significant, and Ewers's battle group was ordered to proceed to Bonn.

General Kortzfleisch was angry and frustrated. He had been given a job to do but not the means to do it. "*Herr Feldmarschall*," the customarily placid Kortzfleisch shouted into the telephone, "I feel obligated to point out that this order will be of decisive consequence in the war."

Depressed and resigned to a major catastrophe at Remagen, General Kortzfleisch and Colonel Schulz drove on south toward the bridgehead. Five miles north of the site of the American crossing, their car was waved down by a tall, disheveled German officer who identified himself as Major Hans Scheller and said that he had been in command at Remagen. He quickly outlined the situation and said that he had ridden off on a bicycle to seek reinforcements. Scheller stressed that the Americans were still weak on the east bank and that an immediate attack could wipe them out. Kortzfleisch's brow furrowed with anguish. There was nothing he could do about the situation.

Courtney Hodges's First Army headquarters at Spa, Belgium, had refused for two hours to approve Bill Hoge's crossing of the Rhine, but at dusk the general returned. The indecision vanished. In his soft-spoken manner, Hodges promptly ordered that "everything available be pushed across the bridge." Here was a golden opportunity to get at the vitals of Nazi Germany.

General Hodges put in a call to his boss, Omar Bradley, at Namur. "Brad," he observed in his customary casual way, "we've gotten a bridge."

"A bridge. You mean you've got a bridge intact on the Rhine?"

"Leonard nabbed the one at Remagen."

"Hot dog, Courtney, this will bust the German wide open! Are you getting the stuff across?"

Hodges said that he was sending the 9th and 78th Infantry Divisions across and asked for approval for the 99th Infantry Division to join them.

"Shove everything you can across, Courtney," Bradley stressed in his high-pitched voice. "It will probably take the other fellow a couple of days to pull enough stuff together to hit you."[1]

By happenstance, Major General Harold "Pink" Bull, Eisenhower's G-3 (operations officer) arrived at 12th Army Group headquarters in the Château de Namur that night. Bull, small and reddish-haired, had long been one of Omar Bradley's closest friends. The SHAEF general no sooner walked through the massive door of the medieval château than a staff officer rushed up to him and blurted excitedly, "Have you heard the great news, general?" He told Bull about the bridge.[2]

Pink Bull's reactions were mixed. He too was thrilled with the Ludendorff exploit, but knew that SHAEF's master plan called for Field Marshal Montgomery to make the main thrust over the Rhine in two weeks. "Well," Bull thought, "I'll hear all about it from Brad at dinner."

Since getting the electrifying news from Courtney Hodges a short time before Bull arrived at Namur, Omar Bradley had experienced one of the most buoyant periods of his life. For weeks Montgomery had been preparing to launch Operation Plunder, 21st Army Group's massive assault over the Rhine, and now American troops had stolen the spotlight from the British field marshal. Bradley was still simmering over Montgomery's press conference following the Battle of the Bulge. More importantly, General Bradley was euphoric over the enormous tactical opportunities resulting from the seizure of the bridge. If the Remagen bridgehead could be rapidly exploited, Bradley's long-held concept of a double-thrust into the Third Reich would be realized. But Bradley knew that gaining General Eisenhower's approval for this "right hook" would be like pulling teeth. The sacrosanct Master Strategic Plan called for Montgomery to make a powerful single thrust into Germany while the U.S. First and Third Armies maintained an "aggressive defense" in the center and south.

Bradley knew that a sudden change of strategic design based on the unexpected seizure of Ludendorff Bridge would bring abuse down on Eisenhower from Churchill, Brooke, and Montgomery. The supreme commander would be accused by the British leaders of a breach of faith in contravention of the one-thrust strategic concept adopted by the Combined Chiefs of Staff at Malta. Yet the psychological impact of the Remagen bridgehead was enormous. The Allied world would not stand for the failure to exploit this golden opportunity.

Mulling over the electrifying developments at the Ludendorff Bridge, Omar Bradley collaborated with Courtney Hodges to get the right-hook concept adopted: Hodges would get his army so deeply committed on the far side of the Rhine that SHAEF could not possibly pull the Americans back. Bradley and Hodges, both low-key straight shooters, were the least likely candidates in the Allied command to connive to achieve a goal. But, they agreed, unique circumstances obligated them to resort to this strategem.

At dinner that night in the towering old castle at Namur, Pink Bull could hardly eat. His mind was awhirl with thoughts—and problems—concerning the bridge. Curiously, Bull mused, his old friend Omar had not even mentioned the Ludendorff Bridge coup of a few hours earlier. Another factor troubled Bull as he tried to chew his food: it appeared that Bradley had not even informed General Eisenhower of the development. From Bradley's point of view, there was good reason to "keep mum" as long as possible before informing his boss, Eisenhower: with every minute that went by, Courtney Hodges was pouring more troops and guns over the Ludendorff.

Bradley was certain that Pink Bull would support him. But when Bradley took Bull to his war room after the quiet dinner and brought up the Ludendorff Bridge shocker for the first time, Eisenhower's operations chief was adamant in opposing its exploitation.

Grabbing the bridge was an heroic feat, Bull observed, but there was little that could be done with it due to the rugged, hostile terrain. "You're not going anywhere down there," Bull told his friend. "Besides, it doesn't fit into the overall plan." As Bradley grew hotter under the collar, Bull added, "Ike's heart is in your sector, but right now his mind is up north [in Montgomery's zone]."[3]

At this point the normally mild-mannered Bradley was thoroughly angry. Bull had become so totally inflexible, or so pro-British, or both, that he was blinded to the opportunities at Remagen, Bradley felt. "What in the hell do you want us to do, pull back and blow it up?" he shouted.

Pink Bull was shocked by his friend's piercing tone. He could see nothing wrong with Eisenhower's operations chief pointing out to a commander the complexities of the larger situation. And why, he wondered, was the 12th Army Group leader trying to gain permission from him, SHAEF's operations officer, to push four or five divisions over the Ludendorff? Eisenhower himself was the one to make such a far-reaching decision.

Now General Bull was sure of it: more than two hours after Bradley received the startling news from Hodges, Eisenhower had not been informed.

It was just past 8:00 P.M. when a relaxed and jovial supreme commander sat down to dinner at his living quarters at Rheims. Eisenhower had reason to be in good spirits. His attacks up and down the lengthy front had been surging ahead, and Montgomery was steadily hauling up supplies, ammunition, and troops for Plunder, which would jump off in two weeks. Eisenhower's guests included Lieutenant General Frederick Morgan, the unheralded British planning genius who had developed the concept and early design for Overlord, the monumental invasion of Normandy. Ike's American airborne commanders, Major Generals Matthew B. Ridgeway, James M. Gavin, Maxwell D. Taylor, and William M. "Bud" Miley, were also in attendance.

Midway through the meal an aide bustled into the dining room. "General Bradley's on the phone, sir," the officer told Eisenhower. "Says it's highly important."

As the supreme commander began talking on a phone in the same room, the assembled generals picked at their food and eavesdropped shamelessly. Clearly, something big had occurred, and Bradley was breaking startling news.

"Brad, that's wonderful!" they heard a beaming Eisenhower say. "How much have you got in the vicinity that you can throw across the river?"

Bradley said he had four divisions nearby, and wanted to send them over the Rhine on the Ludendorff Bridge.

"Sure, get right on across," Eisenhower said without hesitation. "It's the best break we've had. We expect to have that many divisions tied up at Cologne, and now those are free. Go ahead and shove over at least five divisions immediately, and anything else that is necessary to make certain of our hold."

"That's precisely what I wanted to do," Bradley observed, "but the question has been raised here"—he was staring fixedly at old pal Pink Bull—"about conflict with your plans."[4]

Ike's nearby guests had given up any pretext of disinterest in the telephone conversation. They heard the supreme commander shout into the phone, "To hell with the goddamned planners. I'll give you everything we have to hold that bridgehead. We'll make good use of it even if the terrain isn't too good."

Eisenhower, jubilant, hung up and returned to the table, a broad smile across his face. Here, in fact, was just what was needed: a Rhine bridgehead which would draw off German forces from Field Marshal Montgomery's maximum effort in the north.[5]

Bradley, in the ongoing cat-and-mouse game between American and British generals, had other ideas.

13

Hitler Thirsts for Scapegoats

Chaos reigned at both ends of the Ludendorff Bridge a few hours after Sergeant Alex Drabnik had been first across. Now it was dark, and pouring rain saturated the clothing of fighting men on both sides. American engineers were at work planking parts of the bridge and filling in the huge crater that had been blown at the western entrance to the bridge.

Halfway to the top of 660-foot-high Erpeler Ley, all three companies of Major Deevers's 27th Armored Infantry were clinging to the slippery ground as German gunners above fired on them with automatic weapons. Tanks, trucks, artillery, jeeps, ambulances, and an assortment of other vehicles were jammed together at the bridge entrance, waiting for repairs to be made. Milling among all this rolling stock were hundreds of foot soldiers waiting for their turn to cross.

About a block from the bridge entrance, Lieutenant Colonel Leonard Engeman, the task force commander, was solemn-faced as he talked with his commanders. He had been told by his engineering officer, who had been on the bridge during the assault, that the structure would hold tanks. Engeman was not so sure. But he told his tankers, "We've got to try to get across."

It was decided that engineers would string white tape along the center of the span to guide the tracked war wagons in the blackness. Then, on reaching the far bank, Engeman's tanks were to coil up until dawn, then attack. It was nearly midnight when the repair work was completed and the first platoon of new Pershing tanks began to inch across the bridge. The roar of tank motors and grinding of treads echoed up and down the Rhine gorge for miles and ricocheted off the cliff to which Deevers's rain-soaked riflemen were clinging. It was so black that the drivers lost sight of the white tape only a few feet below them. Despite the commotion, not a shot was fired at them.

Led by Lieutenant C. Windsor Miller of Washington, D.C., the tank platoon halted briefly on the far side as Miller peered intently through the blackness trying to locate the riflemen who were to escort the armored unit. None were there. So Miller turned north and headed slowly along the road running beside the Rhine. Suddenly machine guns on all sides opened up against the Pershings. Miller's men fired back blindly.

Shouts rang out in the darkness: "*Kamerad! Kamerad!* [Surrender!]" Miller could not make out whether the Germans wanted to surrender themselves or were calling for the Americans to capitulate to them. In desperation, Miller radioed Colonel Engeman back on the other side of the bridge: "Send some goddamned infantry."

Engeman snapped back, "You will hold that position until the last tank is shot out from under you." Engeman had no infantry to send.

Hearing the heavy firing in the blackness below, Deevers's drenched and exhausted men 300 feet up the cliff became increasingly nervous. A rumor spread among the infantrymen: "We're pulling back to the other side of the river!" Soldiers began scrambling down the face of the cliff. There was near panic as almost a third of those on Erpeler Ley raced back over the bridge to the west side, bowling over startled engineers working on the span.

It was nearing 4:30 A.M. when a battalion of the 47th Infantry Regiment of Lou Craig's veteran 9th Infantry Division edged up to the entrance of Ludendorff Bridge and prepared to cross. Commanded by 25-year-old Lieutenant Colonel Lewis Maness, this infantry battalion was the first of General Hodges's rush reinforcements to reach Remagen. Maness and his men were "out on their feet" after one of the 9th Infantry Division's most grueling days.

At 8:05 P.M. the night before, Colonel George W. Smythe, who had led the 47th Infantry Regiment since June 1943, had received an urgent message from III Corps. Effective immediately, Smythe's regiment, known as the Raiders, was to rush to Remagen and cross the Ludendorff Bridge. Smythe had been elated over the news of the span's seizure; his regiment had been attacking all day and had reached the Raider objective on the Rhine, 25 miles north of Remagen. There was no time for route reconnaissance, however. Smythe's drained men simply climbed into trucks as they arrived and set out for the south.

Young Colonel Maness's battalion was in the lead and had to detruck eight miles short of Ludendorff Bridge and make its way on foot over a narrow, winding road clogged with tanks and half-tracks. The men, buffeted by cold winds, slipped and slithered in the rainy blackness. Curses rang out as men repeatedly fell on the hard road surface.

Reaching the entrance to the bridge, and ready to lead his men across, a higher American commander sought to encourage Colonel Maness. "You'll find nothing on the other side but demoralization," Maness was told. The young battalion commander wondered which side was demoralized over there—American or German.

"Okay," Maness called out, "let's get the hell on over as quickly as possible!" Once onto the bridge, the Raiders needed no exhortation for speed. All 700 men were soon on the far side.[1]

The black sky had begun to lighten to gray as foot soldiers of General Edwin Parker's 78th Infantry Division began rushing over the structure, trying to keep from staring at the angry water far below. On the far side the leading elements encountered 97 German engineers who were headed for the bridge carrying a ton and a half of explosives. A fierce firefight broke out, and the Americans killed or captured the entire enemy force before it could blow up the bridge.

It was nearing 8:00 A.M. when a jeep carrying General Bill Hoge and his operations officer, Major Ben Cothran, drove across the Ludendorff. Maybe they were hearing things, they mused, but once or twice it sounded as though the huge span was creaking. With machine-gun fire chattering angrily in the distance, Hoge's jeep turned left and raced to nearby Erpel, where the combat command leader set up headquarters.

Shortly after dawn, Chaplain Gibble, who had calmly taken movies of the Ludendorff assault the previous afternoon, set up a field altar near the tunnel entrance. "Business" was poor. Most of Gibble's potential "customers" were involved in other pursuits. So the energetic military padre throught he should do something worthwhile. He hopped into a jeep with his aide at the wheel and drove south along the scenic highway to Linz. There the city fathers eagerly surrendered the town to Captain Gibble, making him perhaps the only chaplain to personally seize an enemy community. When five Pershings clanked up to Linz a few hours later, Chaplain Gibble was out to greet the tank commander.

On the German home front, as well as among the armed forces, the crossing of the Rhine would come as a crushing blow to faltering morale. Berlin was silent for more than 24 hours. But the Allied-controlled Radio Luxembourg made certain that the news reached every part of the Third Reich: "This is the greatest blow against the German army. The final phase of the war has begun. . . ."

Late in the morning of March 8, American antiaircraft gunners around the bridge opened up with a tremendous barrage. Ten Luftwaffe fighter-bombers were headed for the crucial span. The war planes were driven off without scoring a hit on the bridge, but long-range German guns had meanwhile gone into action. Shells began to whistle around Ludendorff, killing and wounding many Americans in the traffic clog at the entrance on the west side.

The occasional incoming shells did not halt the steady stream of First Army reinforcements moving across the bridge. Resistance on the far side had been sporadic, and by 4:00 P.M., 8,000 of Courtney Hodges's soldiers were across the Rhine.

While the free world was being electrified by the coup, other First

Army forces were also closing on the Rhine on March 7. Elements of General Lou Craig's 9th Infantry Division and Major General Clift Andrus's 1st Infantry Division (known as the Big Red One) were converging on the large city of Bonn, north of Remagen. On the second day of an attack on Bad Godesberg, a Rhineland peacetime resort, a 9th Division battalion commanded by Lieutenant Colonel Frank L. Gunn was just outside the city, which was being defended by elements of the 3rd Parachute Division. The Germans were part of the Bonn defensive ring.

In heavy fighting, Gunn's men had overcome all opposition except for a battery of 88-millimeter flak guns which was in position along a river and adjacent to a hothouse. Believing the enemy force might surrender, Colonel Gunn asked for a volunteer officer to accompany a German civilian to the enemy commander in the town. Lieutenant Mike Wolfson agreed to go.

Gunn instructed the lieutenant, "Tell him that if he doesn't surrender by sixteen-twenty [4:20 P.M.] we're going to blow hell out of them with artillery and send our infantry in to wipe them out."[2]

Climbing into a jeep with a driver, a runner, and a civilian interpreter, Wolfson headed for the German lines, where he was taken into Bad Godesberg to the captain in command of the defending force. Wolfson delivered the ultimatum, but the German officer was unimpressed.

"We have a strongly fortified garrison," he pointed out. "We are heavily armed and have four eighty-eights aimed at all of your possible approach routes."

"But your situation is hopeless," Wolfson said.

"It would be dishonorable for me to surrender."

The American lieutenant spoke of the "many innocent German civilians" who would likely be killed or wounded as a consequence of the German garrison's refusal to capitulate.

"In that case, tell your Colonel Gunn that we should be given time to disengage and regroup in another sector," the captain said.

The debate continued. Wolfson was growing increasingly nervous. The deadline of 4:20 P.M. was approaching, and American artillery would start pounding Bad Godesberg and German positions at that time, presuming that the American delegation was "lost."

The clock ticked on . . . 4:04 P.M. . . . 4:12 . . . 4:16.

Suddenly, without a word or change of expression, the German captain unsnapped his holster and pulled out a Luger. Wolfson swallowed hard. This was the end. Holding the pistol in his hand for several moments, the officer turned it over to the American lieutenant. Wolfson sought to mask his enormous relief.

Formed into two columns with the captain in the lead, the German defenders, minus their weapons, filed out of Bad Godesberg just as the promised artillery barrage began pounding the town. Among the POWs seized in the fight for Bad Godesberg was Major General Richard Schimpf, the tall, tough commander of the 3rd Parachute Division.

North of Lou Craig's 9th Infantry Division, elements of Clift Andrus's Big Red One overcame stiff resistance and fought their way into the streets of Bonn by late afternoon of March 8. As heavy fighting erupted in the old town, General Andrus sent down an order: seize the Bonn bridge over the Rhine.

The Bonn bridge was the only escape route for Germans trapped in the old city of 140,000, which in peacetime was noted chiefly as the seat of famed Bonn University and the birthplace of Ludwig van Beethoven. Founded by the Romans as a fortified encampment on the Rhine, it had been known as *Bonna* or *Castra Bonnensia*, and there the Romans had maintained their position on the Rhine against restless Germanic tribes on the east bank.

Now, centuries later, German prisoners reported that they had seen the bridge prepared for demolition. Most were astonished that the span had not already been blown. At 9:15 P.M., with machine-gun and small-arms fire rattling all over town, an enormous roar engulfed the old city, and the Bonn bridge lifted into the sky before falling into the dark waters of the Rhine.

Blowing the large structure came as a deep relief for a young German captain of the 6th Engineer Regiment who had not slept for three days and nights worrying over whether he would be able to destroy the bridge at the needed moment. The penalty for failing to do so execution by an SS firing squad.

The blowing of the bridge greatly diminished resistance in Bonn, so the biggest headache for the Big Red One was rounding up the hundreds of German soldiers and members of the Volkssturm who had had enough and were milling around the city, some in civilian clothes. More than 200 Wehrmacht deserters, some of whom had been hiding out for a month, were flushed out of civilian cellars. Most had been in the Bonn area on furloughs, and instead of returning to their units had decided to hole up until the Americans arrived.

South of Remagen on the ninth, nervous German commanders, aware that George Patton's Third Army was slashing toward the Rhine, blew up the massive highway bridge at Koblenz and a railroad span near Andernach. Patton's hard-riding tankers were not yet in sight of the historic river barrier, but the Wehrmacht was making certain that there was no repetition of the catastrophe at Ludendorff Bridge.

In Berlin on March 8, Adolf Hitler flew into a rage on being told by his principal military confidant, Colonel General Alfred Jodl, that the Rhine River had been breached at Remagen. Aides whispered that this was the most violent outburst since the bomb plot the previous July, for which hundreds of military men and civilians, many only suspects, had paid before firing squads or at the end of ropes.

Red-faced and trembling with fury, Hitler shouted that he had been betrayed once again and vowed vengeance upon those responsible for the failure to blow up Ludendorff Bridge. His vengeance would take many forms up and down the chain of command, but he started by summarily sacking the old Prussian, Field Marshal Gerd von Rundstedt, whom Hitler had branded as a "retreat general." Then he placed an urgent phone call to Italy, and instructed Field Marshal Albrecht Kesselring, the commander in the Mediterranean, to report to Berlin at once. Known to the Allies as "Smiling Al," due to the fact that he was nearly always grinning when his picture appeared in German publications, Kesselring would replace the "retreat general" along the Rhine. Hitler overlooked the fact that Kesselring had been retreating steadily for more than two years.

Hitler then contacted a hulking, scar-faced SS officer to whom he had always turned with crucial assignments. He was Standarten-führer (Colonel) Otto Skorzeny, an Austrian who had electrified the world in 1943 with a daring rescue of Italian dictator Benito Mussolini from a mountaintop where he had been held by partisans for trial and probable execution. In the Battle of the Bulge the previous December, Colonel Skorzeny had led a group of English-speaking German soldiers, wearing American uniforms and driving American vehicles, behind General Bradley's lines. The ruse had created panic in the American rear and spread hysteria among civilian populations in Allied-controlled countries of Europe. The Allied press at that time had pinned a terrifying nickname on this German commando leader said to be a hired assassin—The Most Dangerous Man in Europe.

When Skorzeny arrived in Berlin in response to Hitler's urgent summons, the Führer had retired for the night. But General Jodl related to him what Hitler wanted—take a group of specially trained frogmen and destroy Ludendorff Bridge.

For perhaps the first time in the war, Skorzeny hesitated. He pointed out to Jodl that the Rhine River temperature would be near freezing, the current swift; and since the American bridgehead had been extended for a considerable distance up and down the river it would require a near miracle for his team to get the job done. But he promised Jodl he would immediately rush his best men to Remagen.

Meanwhile, American troops, tanks, and supplies continued across the Ludendorff Bridge, and German artillery, which had pulled back over this same span only 36 hours previously, began to pound the structure and the western approaches. The huge iron bridge had suddenly become one of the hottest spots along the western front. From his luxurious estate north of Berlin Reichsmarshall Hermann Göring ordered his Luftwaffe to attack the bridge with all the strength it could muster.

The perilous job of directing traffic at the bridge fell to the men of the 9th Infantry Division Military Police Platoon. These soldiers, with MP arm bands and billy clubs, were often the butt of combat men's wrath in rear areas, but here they gained wide respect among the infantrymen and tankers—and suffered many casualties as the

153

Germans stepped up their shelling. As one MP fell, another dash
up to take his place. Under the division provost marshal, Major Cl
Thurston, the MPs stuck to their posts for six days and nights
guide First Army reinforcements across.

Vehicles were hit and men wounded and soldiers scattered *
cover. But the MP at Ludendorff Bridge could not take cover. Wh
drivers of vehicles in convoy jumped out to seek shelter from *
shelling, the MPs forced them back to the abandoned vehicles
drove the vehicles off the bridge themselves to keep the traffic mo
ing.

If the Germans on the east side of the bridge had been in disarr
during the first 24 hours of the American crossing, Courtn
Hodges's forces there were little better off. The rapidly expandi
bridgehead had created organizational and communications proble
for the Americans, so Hodges quickly reorganized his command
Remagen. General Lou Craig of the 9th Infantry Division was put
charge of all operations on the east side of the Rhine.

It was nearing midnight on March 8 when Craig and his driv
inched their jeep toward the entrance to Ludendorff Bridge to cross
the far bank. Craig recalled the old expression, "It's so black you ca
see your hand in front of your face." He instinctively put his hand
inches from his nose. He could not see the hand.

Craig got out of the jeep and crawled up onto the hood, lying bell
down with his face hanging over the front—an undignified positi
for a general, but a pragmatic one. The driver would move ahead
few feet, then Craig would call out instructions. As his driver edge
along slowly, the 9th Division commander tried to feel the brid
flooring, and periodically thrust his arms out to each side to tell if t
jeep was nearing the railings and a possible plunge into the water 8
feet below. He offered up a silent prayer that no vehicle was comin
in the opposite direction.

Some 350 miles to the east of Ludendorff Bridge at noon on March '
Albert Kesselring strode briskly into Adolf Hitler's office in Berli
Hitler, solemn-faced and drawn-looking, wasted no time with the cu
tomary amenities with which he greeted those of field marshal ran

The disaster at Remagen dictated a change in command at the top, Hitler declared.

"Only a younger and more active commander . . . who enjoys the confidence of his men in the line, can perhaps still restore the situation in the West," he told Kesselring. "I have confidence that you will do everything humanly possible."[3]

Hitler, visibly shaken by the startling turn of events at Remagen, declared that the Ludendorff Bridge was now the most important locale on the western front. It was vital to restore the situation there, and he was positive that it could be done. Kesselring, impressed by the grasp of detail Hitler evidenced, departed at once for Ziegenberg, near Bad Nauheim, to take over his new command.

Hitler's obsessive search for officers responsible for the catastrophe at Remagen had deflected him from the realization that, had he followed von Rundstedt's advice and pulled German forces back over the Rhine in orderly fashion, the Ludendorff Bridge seizure would never have occurred. Hitler, in his rage, was thirsty for scapegoats. He sent for Major General Rudolf Hübner, an SS officer and staunch Nazi.

Hübner arrived at the Reich Chancellery and was told by Hitler that he was to head a Flying Special Tribunal West, which was empowered to conduct on-the-spot courts-martial and immediate execution of sentences. The first targets for this mobile inquisition were the "cowards and betrayers" at Ludendorff Bridge.

At the same time Hitler was briefing Kesselring in Berlin, the Remagen situation was under intense discussion at General Eisenhower's advanced headquarters at Rheims. The supreme commander had lost much of his spontaneous ardor over the Ludendorff Bridge coup and decreed that Bradley could send *only* five divisions—some 60,000 men—into the bridgehead, and that they could advance a maximum of only 1,000 yards each day, presumably a sufficient distance to keep the Germans from mining and digging in to the front. When the Americans reached the north-south Bonn-Frankfurt autobahn, nine miles inland, they were to halt until Eisenhower gave his approval to advance.

The supreme commander had grown cautious. Building up the Remagen bridgehead was proving to be more difficult than envisioned.

American commanders were beset with enormous traffic tangles whil[e] trying to funnel supplies for five divisions over a single bridge. A U.S. Navy unit arrived with LCVPs (landing craft vehicular, person nel), and three rafts were put into operation by the 86th Engineer Pontoon Battalion. These were stopgap measures until treadway bridges could be constructed.

Late in the day on March 9, Omar Bradley received a phone cal[l] from General Pink Bull, SHAEF's operations chief. "Ike wants me t[o] tell you that the [Remagen] bridgehead is to be held firmly and devel oped with a view toward an early advance to the southeast. This wil[l] furnish an assist to Plunder [Montgomery's Rhine crossing] withou[t] detracting from the effort already allocated to Montgomery." Bul[l] concluded, "The order will follow in writing."[4]

Bradley could hardly contain his elation. To accommodate the Brit ish, Eisenhower had been careful with his wording. Bradley's thrust would be an "assist" to Montgomery, who would still make the pri mary crossing. But, the 12th Army Group commander was con vinced, in permitting Bradley to launch "an early advance to the southeast" from Remagen, Eisenhower must have recognized that this would develop into a major offensive rivaling Plunder. With Bradley's powerful thrust knifing into Germany, it would be highly unlikely that Hodges's 10 divisions earmarked for Montgomery's "left hook" would ever be assigned to the British commander.

It was a milestone decision. For the first time in the war, Eisenhower had ordered a major alteration in overall strategy without consulting the combined chiefs of staff or Field Marshal Montgomery. Nazi Germany would be administered a knockout punch by a double blow—Bradley's "right hook" and Montgomery's "left hook." It was the strategy Bradley, Hodges, Patton, and other American command ers had long been campaigning for.

14

"Annihilation of Two Armies Is Imminent"

Late on the night of March 9, Field Marshal Albrecht Kesselring arrived at his new headquarters at Ziegenberg, tired from his long day but mildly confident that he could stabilize the situation on the western front and wipe out the American bridgehead at Remagen. Kesselring had been briefed by Hitler only hours before, and the leader's capacity for self-deception had rubbed off. Hitler had made the task sound logical: the Western Allies had suffered enormous losses over months of heavy fighting, so all Kesselring had to do was to hold on long enough for reinforcements to reach the Rhine.

A 59-year-old Bavarian, Kesselring had served in the artillery for 19 years before transferring to the Luftwaffe, where he had learned to fly at age 49. He commanded the German air force in the invasions of Poland, Belgium, and France, and in 1941 was put in command of all Wehrmacht forces in the Mediterranean. The Allies regarded Kesselring as one of the Third Reich's most capable generals.

Kesselring's first act was to pore over a candid report prepared by his capable young chief of staff, Major General Siegfried Westphal. Kesselring had confidence in Westphal, who had been his chief of staff in the Mediterranean, so he was jolted by the unembroidered report. With 55 battle-worn, often skeletal divisions, giving him on the average 63 fighting men for each mile of front, a shortage of artillery and ammunition, and the supply network a shambles, Kesselring was expected to hold off more than 80 full-strength English and American divisions, plus their masses of artillery and thousands of warplanes.

Into the late-night background of gloom and despair that hovered over Ziegenberg on March 9, General Rudolf Hübner, an ardent Hitler disciple, descended with two aides on Kesselring's headquarters. As the field marshal listened in astonishment, Hübner, who had had no legal training, outlined his role as head of Hitler's mobile court-martial tribunal, empowered to prosecute on the spot any German soldiers in the west, regardless of rank, thought guilty of "cowardice and betrayal."

Kesselring was revolted by the concept of a circuit-riding kangaroo court. He protested heatedly that such procedures would crush whatever morale remained in the Wehrmacht. His pleadings were in vain. Snapping that he had "much more important matters to attend to" than searching for scapegoats, Kesselring stalked out of the room.

At mid-afternoon the following day, March 10, General Hodges jeeped to Remagen for the first time. He found the scene quiet. The bridge was cleared of traffic and the First Army commander's vehicle sped across the span with the driver's foot pressing down hard on the accelerator. Reaching General Craig's CP, Hodges was rendered a Wehrmacht salute—a salvo of shells bracketed the command post, pulverizing an outdoor latrine.

Hodges learned from Craig that in excess of 21,000 men were in the bridgehead and that Craig's 9th and Parker's 78th Infantry Divisions were attacking and gaining their allotted 1,000 yards per day over rugged terrain against sporadic resistance. The 99th Infantry Division was preparing to cross the Rhine and would be operational on the east bank the following day.

General Hodges was not impressed by the way things were being handled in the bridgehead. In the less than one month John Millikin's III Corps had been in Hodges's First Army, there had been growing friction between the two men. Now Hodges was faulting Millikin for what he termed "lack of progress" in the bridgehead, even though gains had been restricted to 1,000 yards daily. Millikin, whose corps had been transferred from Patton's army, was a rangy cavalrymen who thought and acted as a cavalryman, with dash and daring. His top aides were also cavalrymen. Hodges and his chief staff officers were all infantrymen. Hodges may have thought Millikin too brash and impetuous—a Patton type—and for his part the III Corps leader may have considered infantryman Hodges too cautious and plodding. The result of the friction was that Hodges, driving back to his headquarters, made up his mind that John Millikin would have to go.[1]

On the following day, March 11, General Rudolf Hübner launched his first court-martial in a large farmhouse some 28 miles east of the Rhine. Seated on a rickety divan, the three colonel-ranked judges listened impassively as the legal officer of Model's Army Group B presented the "evidence" against the "cowards and betrayers" at Ludendorff Bridge. The proceedings were over in a hurry. Hübner played his role to the hilt, shouting at each defendant that he was guilty of cowardice and deserved to be shot.

Captain Willi Bratge, the bridge security guard leader who had "allowed" the Americans to seize Ludendorff Bridge, was sentenced to death in absentia, at the time being ensconced in the relative safety of an American POW camp. A bewildered Major Hans Scheller, browbeaten for an hour by judges and prosecutor alike, mumbled that he was indeed guilty of cowardice and was dragged away and shot. Young Lieutenant Karl Peters admitted that when he had pulled his flak battery back over the bridge, he may have left one gun behind in haste with American tankers nipping at his heels. Peters was found guilty of "high treason" in only a few minutes; he was taken outside and shot. Major Hans Ströbel, who on his own initiative had launched a small band of engineers with dynamite toward the bridge to blow it up a short time after its capture, was also condemned to death, as was Major August Kraft, chief engineer in the Remagen

159

region, who had never been anywhere close to the bridge. The graying Captain Karl Friesenhahn, whose job it was to blow up Ludendorff Bridge, was acquitted in absentia.

Kesselring was ordered to publicize the trials and executions. His message, which went to each unit on the western front, warned that here was the fate that awaited cowards and traitors. "He who does not live in honor dies in shame," the signal concluded.

Forty miles north of Remagen at Cologne, soldiers of General Lightning Joe Collins's VII Corps were poking through the Nippes railroad marshaling yards in a brief lull between battles. The yards were a shambles from the countless visits over the past five years by heavy bombers of the Royal Air Force and the U.S. Army Air Corps. Roundhouses and repair shops were masses of twisted wreckage. Hundreds of freight cars and locomotives lay rusting on broken sidings. Bomb holes and craters laced the once-bustling yards. Through here in the glory days of the Third Reich the Wehrmacht had hustled troops, tanks, vehicles, and supplies. Now it was ghostlike.

Curious items were found in some freight cars which puzzled the GIs. There was an entire carload of famed Cologne water on its way to the front, and another car was loaded with meat for the troops doing the fighting—horse meat. Tankers of the 3rd Armored Division jested for days over which gave off the more repugnant odor—the cologne used by the Feldgrau or the 10-day-old horseflesh.

The dash by Joe Collins's corps to the Rhine from the Roer and the quick capture of the Third Reich's fourth-largest city had attracted less notice in the Allied world than had the electrifying seizure of Ludendorff Bridge. But Collins intended to see that those who had paid the price for the achievement were not forgotten. At a ceremony in Cologne's finely manicured stadium on March 11, General Collins officially raised the Stars and Stripes on a tall flagstaff as representative officers and men of his three assault divisions were drawn up in parade formation. It was almost identical to the ceremony at which the VII Corps leader had presided in the French port of Cherbourg the previous July.

"At this time," the handsome young general declared, "we pause to remember those men who gave their lives so that we might be here. . . ."

And Collins continued to speak. Around him were the spearhead tankers of Maurice Rose's 3rd Armored Division, their faces scrubbed and uniforms cleaned for the occasion. These men had been powder-burned and deaf not too long ago. There were Terrence de la Mesa Allen's Timberwolves of the 104th Infantry Division, the very fighting men who only days before had come charging through this same *Sportplatz* with rifles, Tommy guns, and BARs blazing. Beside them, wearing the golden arrow of the 8th Infantry Division, were the fighting men who had played a major role in the victory.

Overhead, a flight of Thunderbolts wheeled in the gray sky—they, too, were proving instrumental in the drive to the Rhine and Cologne. The band struck up the "Star Spangled Banner," the weather-beaten young fighting men saluted as one, then only silence. The ceremony was over. There was pride, but no exuberance. Cologne was only a desolate way station. Bitter battles lay ahead.

Meanwhile, General George Patton's Third Army had been pressing toward the Rhine through the deep ravines, thick forests, and rugged hills of the Eifel against resistance that ranged from fierce to none, depending upon the resolve of various local German commanders. Patton's devoted staff thought that their chief was "just a little envious" of First Army's dramatic crossing of the Rhine at Remagen, but there was nothing he could do about it at the moment. Third Army was still a long distance from the river.[2]

On March 3, the 5th Infantry Division of Third Army assaulted the Kyll River in the vicinity of Bitburg, and established a bridgehead; two days later Hugh Gaffey's 4th Armored Division was turned loose. Its mission: pass through the bridgehead and drive northeastward 65 miles to the Rhine River north of Koblenz. As Gaffey's tankers jumped off, fog and rain shrouded the thick woods, and fighter-bombers, which customarily flew in support of the armored spearheads, were grounded. Combat Command B, led by Brigadier General Holmes Dager, was met by anti-tank, panzerfaust, and artillery fire, but advanced for 13 miles the first day anyway. The plunge into German lines threw the defenders into confusion. At one point, Dager's tanks got between an artillery battalion headquarters and its four batteries of guns, which had been trying to halt the advancing American tanks.

On the second day of the 4th Armored's plunge toward Koblenz and the Rhine, General of Cavalry Ernst Georg Edwin Count von Rothkirch und Tritt, commander of the LIII Corps, was riding in a staff car in the vicinity of Putzborn. Von Rothkirch, who was in charge of the sector, was working frantically to locate his scattered troops. He spotted some 200 German soldiers lounging about in a field. Furious at their lack of fighting spirit at a time when American tanks and infantry were attacking through the region, the German general headed for the field. Only too late did he discover that the men were prisoners of the Americans.

As guns of the 37th Tank Battalion were aimed in the direction of the German car, it squealed to a halt. Von Rothkirch slowly emerged with hands in the air.

"And where do you think you're going?" Lieutenant Joe Liese asked.

A resigned von Rothkirch replied, "I guess I'm going to the American rear."

"That's a goddamned good guess," Liese said.[3]

With General von Rothkirch's capture, German forces in the sector disintegrated. Here and there the 4th Armored engaged in bitter fights, but 58 hours after jumping off from Bitburg, Gaffey's tankers were on the Rhine above Koblenz.

During the night, German forces streamed south along the Rhine to try to withdraw across the river at the Kronprinz Wilhelm railroad bridge at Urmitz. Tank fire and artillery decimated the fleeing enemy remnants. As a column of German horse-drawn vehicles was passing over the span, the bridge was blown with an enormous roar.

Patton's Third Army at Koblenz on the Rhine now faced southward, to join in wiping out the last Wehrmacht foothold west of the river in a region known to Allied planners as the Saar-Palatinate. This was a huge triangle that had its base along the Rhine, with the two sides meeting at a point 75 miles to the west. Running along the northern leg of the triangle was the formidable Moselle River. The southern leg was faced mainly by Lieutenant General Alexander M. "Sandy" Patch's U.S. Seventh Army.

Included in the 3,000 square miles of enemy-held terrain were the Saar industrial basin, second only to the Ruhr as a source of the Third Reich's war-making muscle. The Allied high command had long had its eye on the Saar-Palatinate. Not only would its seizure deny Hitler vitally needed resources, but numerous sites along the Rhine between Mainz and Mannheim would be ideal for storming the river barrier.

American efforts to crush German forces in the Saar-Palatinate would face not only a desperate enemy but hostile terrain. Swift-flowing rivers such as the Moselle, the Saar, and the Lauter rivers were backed by the Hunsrück Mountains and the heavily forested Lower Vosges Mountains, rising as high as 2,300 feet.

At Third Army headquarters, a staff officer observed aloud that "this is sure damned poor tank country." The remark was overheard by General Patton, who roared, "In Third Army, there's no god-damned such thing as poor tank country!"[4]

General Eisenhower's plan for this mountain region was for Patch's Seventh Army to launch a powerful assault northeastward in the direction of Worms on the Rhine River. Elements of Patton's Third Army would bolt southward across the Moselle and drive to the southeast in the direction of Worms. Patch's and Patton's converging assaults, Eisenhower hoped, would trap a large German force and prevent its escape over the Rhine.

In his forthcoming armored dashes into the Saar-Palatinate, George Patton would have an ally whose existence was known only to him, his chief of staff, and his intelligence officer. That ally was Ultra, the top-secret decoding operation at Station X in Bletchley Park, north of London. When General Gaffey of the 4th Armored made his 65-mile, two-day dash to the Rhine, he bolted through the seam between German army groups B and G, where coordination was weakest. Ultra had intercepted a German signal that pinpointed this juncture between the enemy army groups and, like any brilliant quarterback,

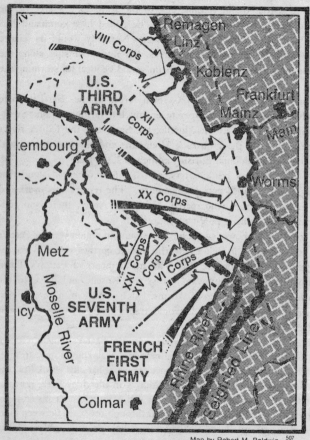

The Saar-Palatinate Triangle

Map by Robert M. Baldwin

507
504

164

One of the scores of Rhine bridges that were blown up in the face of American columns racing for the river. This twisted wreckage was the span leading from Homberg to Duisburg (far side) in General William Simpson's Ninth Army sector. (U.S. Army)

Field Marshal Bernard Montgomery (standing, right) and Lieutenant General Brian Horrocks (left). Horrock's British XXX Corps paced 21st Army group's assault on the Rhine River in Operation Veritable. (Imperial War Museum)

Führer of the Third Reich and Commander-in-Chief of the Wehrmacht, Adolf Hitler. (U.S. Army)

Major General Leland S. Hobbs, 30th Infantry Division. (U.S. Army)

Lieutenant General Sir Miles Dempsey, British Second Army. (U.S. Army)

Major General Maurice Rose, 3rd Armored Division. (U.S. Army)

The high brass confer. From the left: Supreme Commander Dwight D. Eisenhower, Lieutenant General Omar N. Bradley, and Major General J. Lawton Collins. (U.S. Army)

Lieutenant General Courtney H. Hodges. His First Army troops were first over the Rhine. (U.S. Army)

S.S. Obergruppenführer (General) Paul Hausser, Commander, German Army Group G. (U.S. Army)

Lieutenant General George S. Patton, Jr. (U.S. Army)

Feldmarschall Walther Model, Commander, German Army Group B. (National Archives)

Major General Norman D. Cota. Theft of his super-secret decoding machine caused a widespread search. (U.S. Army)

Major General J. Lawton Collins, captor of Cologne. (U.S. Army)

American troops and vehicles pour across the Ludendorff Bridge at Remagen. The towering Erpeler Ley looms at the far end of the Rhine span. (U.S. Army)

A Sherman tank, followed by jeeps and trucks, crosses a pontoon bridge during Patton's drive in the German Saar-Palatinate. These vehicles belong to the 11th Armored Division. (U.S. Army)

A vehicle of Third Army's 4th Armored Division races past a burning German truck in the dash for Worms on the Rhine. (U.S. Army)

Major General Eric L. Bols, commander of the British 6th Airborne Division in Operation Varsity. (Imperial War Museum)

A British paratrooper examines a city limits sign at Hamminkeln, which was bitterly contested by the Germans after the Allied airdrop. (Imperial War Museum)

Major General James M. Gavin, slated to lead the secret parachute assault on Berlin. He was the youngest American two-star general since the Civil War. (U.S. Army)

Feldmarschall Albrecht Kesselring, who replaced von Rundstedt. (U.S. Army)

Colonel James W. Coutts, a leader in the Rhine parachute assault. (Author's collection)

Lieutenant Colonel Edward S. Branigan, Jr., 17th Airborne Division Artillery. (Author's collection)

Corporal Frederick G. Topham, a Canadian paratrooper awarded the Victoria Cross. (Lou Varrone)

Patton sent his best ball carrier, Hugh Gaffey, through the weak point in the opposition's line.

On the day Patton's 4th Armored Division had reached the Rhine south of Koblenz, an Ultra bonanza was being analyzed by Omar Bradley at his 12th Army Group headquarters. It was a complete listing of all German corps and divisional headquarters in Army Group G, which was facing Patton and part of Hodges's First Army. This crucial information was rapidly shuttled on to Patton and Hodges.

While powerful American forces were preparing to strike into the Saar-Palatinate, Field Marshal Kesselring paid his first visit to head-quarters of Army Group G, commanded by the energetic 65-year-old Obergruppenführer (SS General) Paul Hausser. With Hausser were his First and Seventh Army commanders.

General Hausser had distinguished himself in fighting on the Russian front, where he had lost an eye. In Normandy he had been given command of Seventh Army—the first SS officer to achieve that status—when its leader was killed. In France, Hausser had received a serious head wound but rode out of an Allied trap at night while perched on top of a Tiger tank with his Schmeisser machine pistol. In January 1945, Hitler boosted Hausser another notch to command of Army Group B, now something of a hodgepodge of decimated units trying to hold the Saar-Palatinate. Paul Hausser's courage had never been questioned. But now he was laying it on the line to his senior, Kesselring.

"Unless heavy reinforcements can be brought in to strengthen Seventh Army [facing Patton]," Hausser declared, "successive with-drawals behind the Rhine are the best Army Group G can hope to accomplish. Otherwise, envelopment and annihilation of two armies will be imminent."[5]

As Kesselring, without comment, absorbed these gloomy words, General Hans Felber, leader of Seventh Army, took up the cudgel. During the previous month, Felber had initiated a secret "double orders" plot with subordinate leaders to thwart Hitler's edict of not giving up a foot of ground. The strategy of an all-out defense west of the Rhine, Felber insisted, could only result in wholesale German losses, and probably in total destruction.

"The [U.S.] Fourth Armored Division is concentrating near Koblenz," Felber pointed out, "for a quick thrust south that could cut the entire army group off from the Rhine."

Whatever Albrecht Kesselring's true feelings were, he promptly rejected the urgent appeals by two seasoned commanders to pull Army Group G back behind the Rhine while time remained to carry out the maneuver. "The positions have to be held at all costs," the field marshal told his solemn-faced generals.

With Patton attacking toward the southeast and Patch to the northeast, there was danger of the two forces colliding. Therefore, a boundary had been established along the generally east-west flowing Nahe River in the Saar-Palatinate. But Patton had no intention of halting at the Nahe. He confided to his staff that his goal was to seize bridgeheads over the Rhine in the vicinity of Mainz, Oppenheim, and Worms. Reminded by staff officers that these towns were in Patch's zone of advance, Patton said, "To hell with the goddamned boundary."

At 2:45 A.M. on March 13, in a steady drizzle, 31 artillery battalions of Major General Walton H. Walker's XX Corps opened a thunderous barrage on German positions at the apex of the Saar-Palatinate triangle, 75 miles west of the Rhine. This was the opening blow of Operation Undertone, the all-out offensive by Third and Seventh Armies to neutralize the Wehrmacht in the triangle. The Germans resisted fiercely, but on the second day the 10th Armored Division was able to race eastward for 25 miles.

Patton now delivered a second blow to the reeling Wehrmacht. At 2:00 A.M., on the heels of a 30-minute artillery pounding of German positions, elements of Brigadier General Hubert L. Earnest's 90th Infantry Division and Major General S. Leroy "Red" Irwin's 5th Infantry Division paddled assault boats across the Moselle River south of Koblenz. Concealed by darkness and a heavy fog, the crossing was marred only by sporadic bursts of German machine-gun fire. Reaching the far side, the infantrymen pushed ahead, encountering opposition mainly in villages.

Red Irwin, sensing that his foot soldiers had made a breakthrough, ordered trucks to be brought forward to haul his men in a dash to the Rhine in the Mainz-Oppenheim-Worms region, but his boss, Major General Manton S. "Matt" Eddy, leader of XII Corps, overruled the plan.

"Sorry, Red," said the hulking, bespectacled Eddy, "but this is a job for armor. I'm committing Gaffey [4th Armored] in the morning."[6]

Irwin was angry. His men had assaulted the Moselle, and punched a hole in German lines. Now the tankers would reap the headlines.

"But I can be halfway to the objective while the Fourth Armored is getting itself sorted out after it moves across the Moselle," General Irwin fumed.

"Nope, this is a job for armor," Eddy said again.

Early the next day, March 16, Gaffey's 4th Armored crossed the Moselle, passed through the infantry, and dashed southward into the Hunsrück plateau against virtually no resistance. At the narrow Simmer River, bridges were intact, and Gaffey's spearheads pressed on to Bad Kreuznach, a junction of three rail lines and four highways 35 miles southeast of the Moselle jump-off point. Infantrymen of the 90th Division tagged along on Gaffey's left, mopping up bypassed pockets and taking old Rhine towns—Boppard, St. Goar, Bingen.

During the dash of one armored spearhead, Captain Roland Jensen of Portland, Oregon, an artillery officer, was jeeping forward to contact advance units. On the way he passed areas where there had been enemy activity. Jensen dismounted to continue forward on foot. Turning off the road, he saw a bare-headed, bare-footed American soldier alongside a pond, fishing nonchalantly.

"Hey," the captain shouted, "don't you know there's enemy around here?"

"No," came the casual reply. "It's safe enough around here. We've gone clear *past* the Kraut lines."[7]

A battery of field artillery, part of the 4th Armored Division, moved into Heimersheim, the first American troops to stop in the town. The battery commander, a captain, promptly sent for the *Bürgermeister*, or mayor.

"I'll give you precisely thirty minutes to see that every firearm town is turned in," the American stated firmly.

The *Bürgermeister* hurried off to comply. The church bell in t square pealed to alert the 1,900 citizens huddled fearfully in hom and cellars, and a town crier, a ponderous man with a booming voic paced through the streets calling out for citizens to turn in their arm Within 20 minutes, the inhabitants had brought in their weapons.

German forces in front of Patton's slashing tank spearheads were total disarray. Thousands of enemy prisoners clogged the road trudging along unescorted, waving white handkerchiefs. Many aske to be picked up by American vehicles going to the rear. Others soug instructions on where to go. No one wanted to halt to take charge the confused prisoners.

"Keep on going to the rear, goddamn it!" passing Americans wou shout.

Major Enoch J. Scalan of Edwardsville, Illinois, provost marshal Eddy's XII Corps, received an urgent signal to come and take charg of 5,000 German prisoners. The POWs were 25 miles ahead. Scala had only 12 men and two trucks to do the job. He issued a curt orde "Go up and build a cage around them." The 5,000 Germans lounge docilely in a large field while Major Scalan's men fenced them in.

That same day, elements of Eddy's XII Corps driving south fro near Koblenz along the Rhine linked up with armor of Walto Walker's XX Corps moving east from the nose of the Saar-Palatinat triangle. Trapped were remnants of 10 German divisions in th Hunsrück Mountains, victims of Adolf Hitler's inviolate no-with drawal edict.

Meanwhile, Patton turned his attention to seizing Koblenz, whic had been sealed off after Third Army reached the Rhine there. Thre thousand artillery shells rained down on the city at the confluence o the Rhine and the Moselle—one of the projectiles shattering a larg statue of Emperor Wilhelm I. With a peacetime population of 79,000 Koblenz had been an active trading and manufacturing center produc ing ships, pianos, and machinery, as well as Moselle wine. Nov Koblenz was a shambles in the backwash of the war.

It was dusk when the artillery pounding ceased. An America

German tank equipped with a loudspeaker rolled up to the outskirts Koblenz and read a surrender ultimatum to the survivors of the garrison. There was no reply. At 3:00 A.M. elements of Middleton's Corps slipped into the first houses and were met only by weak mortar and machine-gun fire. A few hours later Koblenz was in American hands.

Tattered remnants of 14 different German units were rounded up, many of them elderly Volkssturm militia. Most of the 500 POWs were furious at other German troops who had fled across the Rhine, blown the bridges after them, and left the Koblenz garrison to be trapped by the Americans.

One of the German prisoners was a slight 15-year-old boy. He was shaking violently as tears poured down his cheeks. The youth said he had been drafted two weeks before, handed a gun, and sent to help defend Koblenz. His officers had told him that if he was captured by the Americans they would shoot him. Now, he feared, his time had come.

15

Patton: "Holding
the Hun by
the Nose"

Shortly before dawn on March 15, an enormous roar echoed along 70-mile front of Alexander Patch's Seventh Army on the southern le of the Saar-Palatinate triangle. Hundreds of big guns and mortars— anything that would shoot—began pounding positions of General o Infantry Hermann Foertsch's First Army. Jumping off at daylight on the heels of the heavy bombardment, Patch's infantrymen and tank ers, with elements of General Jean de Lattre de Tassigny's Frencl First Army on their right, headed for their primary objective—th ancient Rhine city of Worms.

General Patch, an infantryman who had seen heavy combat on Guadalcanal in the Pacific months before, had led Seventh Army i the invasion of southern France in August of 1944, followed by a swift advance northward toward the German border. Patch's army had numerous inexperienced units in its ranks, but retained a hard core of tough divisions battle-tested in North Africa, Sicily, Italy, and France.

In the Allied scheme of things, it was not considered crucial that Seventh Army make major territorial gains. As long as it was pinning down large numbers of Germans in heavy fighting ("holding the Hun by the nose," as George Patton described the situation), Patton's columns coming rapidly down from the north could get in behind the enemy facing Patch and trap the bulk of two Wehrmacht armies.

Major General Wade H. Haislip's XV Corps, beefed up to six divisions, spearheaded the assault. For two days Haislip's foot soldiers slogged ahead, bloody yard by bloody yard, running into an especially hard challenge at Bitche ("Bitch" to the grim-faced troops assaulting the stronghold). Surrounded by thick-walled fortresses of the old French Maginot Line, Bitche had been seized from the Germans in bitter and costly fighting the previous December, then had to be given up during the German surprise counteroffensive, which came to be known as the Battle of the Bulge.

At many locales in front of the attacking U.S. Seventh Army, the fighting was savage. The Americans were raked by dug-in automatic weapons with excellent fields of fire. Anti-personnel and anti-tank mines were thickly sown. In the concrete pillboxes there, two or three determined men with a machine gun could hold up for hours the advance of an entire company, or even a battalion.

The terrain facing the Seventh Army attack was rugged, and therefore ideal for the defenders. At the very start of Patch's assault, a regiment of the 45th (Thunderbird) Infantry Division, commanded by 38-year-old Major General Robert T. Frederick, was faced with a water obstacle. That regiment had to fight its way across the Blies River, which it did shortly after jump-off. It pressed forward rapidly, penetrating the German main line of resistance on the far shore, and by nightfall had advanced three miles.[1]

Major General John E. Dahlquist's veteran 36th Infantry Division, whose men had initially come from the Texas National Guard, ran into bloody fighting trying to hack its way through heavily fortified hills and villages. Private First Class Silvestre Herrera of Dahlquist's

142nd Infantry Regiment charged a German strong point that was holding up his battalion's advance and captured eight Feldgrau. Minutes later Herrera and his platoon were pinned down again by heavy streams of fire from German machine guns. In front of the prostrate platoon was an enemy minefield.

Sizing up the situation, Herrera called to his comrades, "Keep me covered." While they opened fire on the German machine-gun positions, Herrera dashed forward into the minefield as the enemy gunners concentrated their fire on him. Halfway to the German guns, Herrera stepped on a mine. His comrades watched in anguish as he was engulfed in a cloud of black smoke.

With bullets thudding into the ground around him, Herrera glanced down at his leg. Blood was spurting from two stumps—the exploding mine had blown off both feet. Gritting his teeth and grimacing, the young soldier pulled himself arduously over the ground to where his rifle had landed. He opened fire at the Germans manning the machine-gun nest, shooting with such accuracy that the enemy crew ducked down and ceased firing. Herrera's comrades took advantage of the enemy lull to edge around to the flank, from where they charged the machine gun and killed its crew. Only then did Herrera lapse into unconsciousness. (He survived and went on to be decorated for his courage.)

On the afternoon of March 16, General Eisenhower's personal airplane, a converted C-47, landed at an airport outside Luxembourg City. As the supreme commander hopped out he was greeted by a beaming George Patton. The Third Army leader, in high spirits, bundled Eisenhower off to his headquarters.

At dinner that night, wine flowed and the talk was boisterous, often profane. Eisenhower was buoyed by the news that Gaffey's armor was racing for the boundary with Seventh Army at the Nahe River. Patton too was elated, knowing secretly that he had no intention of halting at the dividing line. Ebullient as always, Patton spared no effort in rolling out the red carpet: four attractive Red Cross Clubmobile girls, whose job normally was to dole out coffee and doughnuts to fighting men back from the front, were spruced up for the occasion with dress uniforms, white scarves and gloves, and glossy dress shoes—at the "suggestion" of George Patton himself.

Flattered to be in the presence of several of the most exalted names in the Allied world, the young Red Cross women were particularly enchanted with Patton—his humor, his arrogance, the manner in which he alternated between the most vulgar expressions and cultivated speech without missing a beat.

Far to the north in battered Cologne, General Joe Collins had been growing increasingly restless over VII Corps's inactivity for the past two weeks. SHAEF's master plan did not call for a Rhine assault in the Cologne region. The dramatic coup at Ludendorff Bridge had changed all that. Collins was ordered to sideslip his corps to the south and to cross the Rhine on III Corps's floating bridges. When Collins had his headquarters established east of the Rhine, he was to command the Remagen bridgehead's northern half.

As Collins's mud-spattered jeep edged over a Rhine River pontoon bridge on March 16, memories churned in his mind. In 1919, as a youthful lieutenant on duty in Koblenz with the Army of Occupation, Collins had often crossed the Rhine in this vicinity. On this night, with other duties dispensed with for the time being, the corps commander took up a pen and dashed off a letter to his wife, the former Gladys Easterbrook, whom Collins had married at the Kaiserin Augusta Palace chapel in Koblenz in 1921:

> The river was as green and as swift as I remember it, and its terraced vine-clad hills and placid villages as charming as ever. I had intended, like most American soldiers, to at least spit in the Rhine, but I was too lost in memories to remember such rudeness. . . . If only they [the German people] had only had sense enough to sit tight on their riverside benches when the neurotic Hitler awoke them to illusions of grandeur![2]

Needing pontoon bridges to support his corps, Collins called in his chief engineer, Colonel Mason J. Young.

"Young, I believe you can put a bridge across the Rhine in twelve hours," Collins declared, knowing that this would establish a record for World War I and the present war which the corps leader felt would endure for all time. "What kind of a prize do you want me to give you for doing it in less time than that!" Collins had not been pinned with the "Lightning Joe" sobriquet due to his lethargy.

Colonel Young swallowed hard and paused for several seconds. He was being asked to perform a miracle, and his engineers would almost certainly come under artillery fire while at work.

"I'll tell you what I'll do," the corps leader exclaimed. "Throw the bridge up in less than twelve hours and I'll buy beer for the entire engineer battalion!" Collins had 10 hours in mind.

Colonel Young's owl-like face glowed. "I'll accept that challenge," he declared.

"All right," Collins replied. "I'll get the beer if you get me a bridge in less than twelve hours."

Just south of Bonn, the 237th Combat Engineer Battalion, commanded by Lieutenant Colonel Hershel Lynn, assisted by Company E, 23rd Armored Engineer Battalion, and a company of the 238th Combat Engineer Battalion, pitched into their work with the fervor of men with long-parched throats. The bridge, measuring 1,308 feet—longer than four football fields—was completed in 10 hours and 11 minutes. In his mind, a beaming Collins chose to overlook the 11-minute "tardiness" for the goal he had secretly set.

Now Collins was beset with a logistical problem for which his long years of military study had not prepared him. In the disarray of the Rhineland, where could sufficient beer be located to quench the thirst of more than a battalion of combat engineers? Collins dispatched task forces up and down the river on both sides, and the beer was located.

Relishing their construction triumph, the jubilant engineers posted a sign at the bridge:

THE BEER BRIDGE—SHORTCUT TO THE C.B.I.

The CBI was the China-Burma-India theater to which, presumably, many American units would be transferred as a staging area for the invasion of Japan once the Third Reich was crushed.[3]

Posting signs on bridges had become a hallmark along the Rhine.

American reinforcements coming over the Ludendorff Bridge saw a large lettered panel which read:

CROSS THE RHINE WITH DRY FEET.
COURTESY OF THE 9TH ARMORED DIVISION

The Third Reich was shrinking under the Allies' pressure on all sides, yet in his capital city Adolf Hitler was breathing defiance over Radio Berlin. On the tenth anniversary of compulsory military service under the swastika, he alluded to his personal martyrdom and concluded: "We witness in both the east and the west what our people will have to face. Our task is therefore clear; to put up resistance and to wear down our enemies so that, in the end, they will weary of war and will yet be broken!"[4]

Meanwhile, he continued literally to scream at his commanders to destroy the bridge at Remagen. It had become a symbol for the looming destruction of Nazi Germany. The Luftwaffe regularly attacked the span with old Stuka dive bombers and with the revolutionary jets. None had scored a hit, but numerous bombs had exploded nearby. Long-range artillery pounded the river span day and night, a few shells striking the target, others exploding around its foundation.

The Karl howitzer, an enormous 540-millimeter weapon, was set up at the southeast corner of the Remagen bridgehead, some 20 miles from the Ludendorff Bridge. The howitzer itself weighed 132 tons and fired a shell of 4,400 pounds. As it began lobbing projectiles toward the bridge, it sent terror through the Americans around the span. Its huge projectiles approaching sounded like freight trains rushing through tunnels. The bridge, screened by the 660-foot-high Erpeler Ley on the east, was difficult to hit, however, and after pulverizing several houses with its first few rounds, the howitzer had to be removed for repairs.

In his desperation, Adolf Hitler ordered that his *Vergeltungswaffen* (vengeance weapons), the supersonic V-2 rockets, be turned against the Ludendorff Bridge. For eight months these terrifying weapons had been pounding London, creating enormous damage and inflicting thousands of casualties. Now, for the only time in the war, they would be employed in a tactical capacity.

A total of 11 V-2s were fired from bases in the Netherlands. Three landed in the river a short distance from the bridge, five others exploded west of the span, and another, its guidance mechanism gone awry, crashed west of Cologne, well up the Rhine River.

Learning shortly in advance of the intention to "wipe out" the Remagen bridgehead and the Ludendorff Bridge with V-2 rockets, several Wehrmacht battle leaders on the site protested to their military superiors, citing the peril from the relatively inaccurate weapons to thousands of German civilians still in the region. Hitler brushed aside the protests, stating that he was determined to halt the Americans at Remagen with his rockets.

For 239 hours the triple-span Ludendorff Bridge had held together sufficiently to serve its American captors well. But it had taken a terrible beating for most of that time—from German artillery, dive-bombers, V-2 rockets, and Captain Karl Friesenhahn's early explosion on the deck minutes before Lieutenant Timmerman and his men had dashed across. Big American 8-inch howitzers and antiaircraft guns around the bridge had been going off steadily for 11 days and nights, all serving to weaken the structure. For the first four urgent days the bridge had withstood the vibrations and load of tanks, big guns, heavy trucks, and thousands of marching men.

By the morning of March 17, the bridge was no longer of such paramount significance to the Americans, who by now had slung together several pontoon bridges at the site. The Ludendorff was closed for repairs on this day, as it would still be useful in supplying the bridgehead.

At 2:00 P.M. some 200 engineers were laboring on the bridge to strengthen it. Captain Francis Goodman, an engineer supply officer, had been walking back and forth over the span for two hours, and at 2:45 P.M. he stepped off the structure at the west end. To his practiced eye, the repair work seemed to be progressing routinely, and the Ludendorff showed no abnormal symptoms of stress. But the span had but 15 minutes to live.

Lieutenant Colonel Clayton A. Rust, commanding officer of the 276th Engineer Combat Battalion, and an aide were walking toward the east end of the bridge at about 3:00 P.M. when they heard what sounded like a sharp rifle-crack. They froze. Rust looked up anxiously in time to see a girder break loose and dangle from the bridge. Abruptly, another sharp crack from behind was heard. The entire bridge deck started to tremble. Terror surged through all. Engineers and workers dropped tools to run for either end of the bridge—the structure was obviously about to collapse. The entire deck started to shake, and clouds of dust rose from the flooring.

Suddenly, Colonel Rust found himself running uphill as the 500-foot-long center section tilted, then plunged into the current 80 feet below. Moments later Rust realized that he was pinned underwater by a piece of the wreckage. "This is the end," he thought. He seemed in the grip of a huge vise. Rust held his breath as long as he could . . . then his body suddenly broke loose from the trap and he bobbed to the surface, gasping for air. The current carried him downstream to a pontoon bridge, where he was hauled from the Rhine, badly shaken but uninjured.

Another engineer who had been thrown clear of the collapsing bridge paddled laboriously to the Remagen side of the river. Shivering, panting from the exertion, he pulled himself out of the water and edged up a low mound, then turned around and, with hands on hips, surveyed the wreckage. Of Irish descent, the man blurted out, "Now ain't that a hell of a St. Patrick's Day."

Seven men of the 276th Engineer Combat Battalion and the 1058th Bridge Construction and Repair Group were killed; 18 were declared missing and their bodies were never found; and three died later of wounds—a total of 28. About 60 others who were thrown into the water sustained injuries.

A few hours after the Ludendorff Bridge fell, a final ironic note was sounded. Radio Berlin blared out that the "cowards and traitors" responsible for the Rhine disaster had been shot—three majors and a lieutenant.

That same day, Courtney Hodges telephoned his corps com-

mander, John Millikin, to tell him he was to be sacked. "I have som[e]
bad news for you—" Hodges began.

Millikin interrupted him. "Sir, and I have some bad news for you[.]
The railway bridge has just collapsed."[5]

Replacing Millikin as commander of III Corps in the Remage[n]
bridgehead was Major General James A. Van Fleet, a former Wes[t]
Point classmate of both Generals Eisenhower and Bradley. Jim Va[n]
Fleet had gone over the Normandy beaches on D-Day as a colonel i[n]
command of a 4th Infantry Division assault regiment. At 54, he ha[d]
been "old" for his rank, but his performance had been so noteworth[y]
that he rapidly received his first star, then was promoted again t[o]
command the 90th Infantry Division. In nine months Van Fleet ha[d]
risen from colonel to corps commander.

It was 7:00 P.M. on March 17, and several furtive figures, burdene[d]
with peculiar-looking gear, stole through the darkness and slippe[d]
into the rapid current of the Rhine River north of Ludendorff Bridge[.]
These were Colonel Otto Skorzeny's frogmen, who had been ordere[d]
by Hitler to blow up the crucial span. Lieutenant Hans Schreiber[,]
leader of the frogmen, had learned a short time before that the Ludendorff Bridge had already collapsed, but there were still bridges a[t]
Remagen to destroy—pontoon bridges. Schreiber and each of his me[n]
wore a skin-tight rubber suit, rubber foot fins, and carried an apparatus to breathe underwater. About two miles from the Ludendorf[f]
Bridge wreckage, the German swimmers were spotted by alert Americans who had been watching for just such an effort. Suddenly[,]
Schreiber and his men were nearly blinded by powerful searchlights
mounted on tanks. Machine-gun and rifle fire peppered the commandos; several of them were hit. Those who survived paddled for the
shore and were taken prisoners—including their crestfallen leader[,]
Lieutenant Schreiber.

General Omar Bradley, already looking past the Rhine, flew to
Rheims on March 18 to confer with the supreme commander and
push for his right hook out of the Remagen bridgehead, an operation
now code-named Voyage. Bradley got all he asked for—and then

me. Eisenhower agreed to increase the forces in the bridgehead om 5 to 10 divisions by bringing Clarence Huebner's V Corps :ross. And with Patton running wild in the Saar-Palatinate, the su- reme commander gave his approval for Third Army to cross the hine "on the run" near Mainz and drive on north to link up with Iodges's First Army, which would attack to the southeast out of the .emagen bridgehead.

Flying back to his headquarters, Bradley summoned Hodges and atton to meet him in Luxembourg on March 19. There he advised is two army commanders that he had received permission from .isenhower for First and Third armies to play major roles in the orthcoming drive deep into the heart of the Third Reich.

Then Bradley threw a damper over their enthusiasm: Hodges could ot jump off from the bridgehead until *after* March 23. That was the recise day that Field Marshal Montgomery would launch Plunder, ne truly massive crossing of the Rhine which in its scope would be :cond only to the Normandy invasion the previous year.

As Hodges and Patton listened in silence, their boss went on to say omething even more depressing: Patton could not cross the Rhine at Loblenz, which was already in Third Army's hands, but rather vould have to vault the wide river barrier farther south, at a point vhich was still held by the Germans and was more than 10 miles rom Patton's spearhead.

George Patton returned to his headquarters in a foul mood, angry nd frustrated. He was convinced that Montgomery had connived vith SHAEF to throw a halter over Hodges and himself until Opera- ion Plunder had gained the world spotlight. If Patton was going to eat Montgomery over the Rhine, he had four days left in which to each the river in the Mainz-Oppenheim-Worms region, clear the vest bank of Germans, and mount an assault over the water barrier. This task might be militarily impossible, Patton conceded, but he ntended to try it anyhow.

16

"For God's Sake, Tell Them I'm Across!"

In Field Marshal Bernard Montgomery's 21st Army Group sector i
the north, massive preparations were in progress for launchin
Plunder. Field Marshal Kesselring, whose job had been defined b
Adolf Hitler as holding the Rhine until reinforcements arrived, wa
casting increasingly nervous glances at the intense activity i
Montgomery's zone.

If there was any lingering doubt among the Wehrmacht high com
mand as to what the British field marshal had in store for th
beleaguered German forces on the east bank, Sir James Grigg, Grea
Britain's War Secretary, put it to rest. Grigg stood in the House o
Commons in London and announced, "Our forces are getting ready
for a massive crossing of the Rhine."

At Duisburg, north of Cologne, Bill Simpson's Ninth Army patrol
and German patrols were passing each other in the night as they
slipped across the Rhine to the opposite banks to reconnoiter. Germar

ctory workers found themselves suddenly in a ground combat zone
s Simpson's guns on the west bank pounded the industrial plants at
Duisburg. Often the barrages came as employees were reporting to or
aving work. Ninth Army's larger guns brought shells down on the
prawling Krupp munitions works in bomb-battered Essen, some 15
niles back from the Rhine.

Meanwhile, in the Saar-Palatinate to the south, General Felber of
he German Seventh Army, trying to stem George Patton's spear-
eads, and General Foertsch of First Army, which was being cut to
ieces by Sandy Patch's men, had been constantly on the phone to
General Paul Hausser, the one-eyed leader of Army Group G. Felber
nd Foertsch had been pleading for days to pull back behind the
Rhine while there was still time. The alternative, they stressed,
vould be destruction of both German armies on the west bank of the
istoric river barrier.

Hausser bucked the thorny question up to Kesselring, who was too
agy to be seen issuing withdrawal orders. Had he done so, he would
ave been risking Hitler's wrath—and his own neck. Fully aware of
he looming disaster facing Hausser's two armies in the Saar-Palati-
ate, Kesselring on March 18 issued an ambiguous order. For Hitler's
onsumption, he sternly directed that "present positions will be re-
ained." Then he added a key qualification: "An encirclement and
vith it the annihilation of the main body of troops is to be avoided."[1]
How could the Führer fault Kesselring for being on guard against the
ntrapment of his armies?

General Hausser, a wily warrior in his own right, interpreted Kes-
elring's order as the commander in chief west had intended him to.
Hausser directed Felber's and Foertsch's armies to pull back.

In the meantime, General George Patton continued to apply pres-
ure—not only on the Germans but on his own commanders. He
lashed about the countryside from CP to CP, cajoling and browbeat-
ng. "Get the hell going, goddamn it!" echoed through the rafters of
ll his headquarters.

One of those who received the brunt of Patton's wrath daily was
Matt Eddy of XII Corps, whose armored spearheads were running
vild through the Saar. Eddy was liked and highly regarded by the

Third Army commander, but that did not moderate Patton's pressure.

Eddy telephoned a subordinate and told him, "I've never seen th heat as heavy as this. George has been in here goddamning and shou ing about, waving his arms and stomping up and down the room."[2]

On March 18, the hulking, keen, bespectacled Eddy committed h 11th Armored Division. His confidence in this relatively new divisio had been shaken when it had performed poorly in the recent Eife campaign. Its commander had been sacked. Now, under Brigadie General Holmes Dager, Eddy hoped for an improvement. Dager ha distinguished himself as a combat command leader in the 4th A mored Division.

Matt Eddy's hopes were realized—and then some. Dager's tanker plunged 20 miles to the Nahe River at Kirn, and the following da raced another 19 miles to the southeast in the vicinity of Kai serslautern. It was the most spectacular dash of Third Army's Saa Palatinate campaign.

Near Kaiserslautern, a German fired a panzerfaust, the rocket strik ing a tank of the 41st Tank Battalion. The platoon sergeant wa wounded, and other crew members bailed out of the vehicle an scampered for cover. Private First Class Herbert J. Burr remained i the tank and drove it forward alone into the town of Doermosche Rounding a turn, he was suddenly confronted with the muzzle of high-velocity 88-millimeter gun pointing directly at him, with i crew preparing to fire. Unable to both operate the gun on his tank an drive, Burr charged the German gun, expecting an armor-piercin round at any moment. The German gun crew fled as Burr's tan smashed into their battery, crushing it under his treads.

Patton's daring armored thrusts into the disorganized German Sev enth Army were the kind of stuff that generated headlines back home Largely unnoticed by the media was the series of brutal skirmishe that frequently took place. Private First Class Frederick C. Murphy, young medic with the 65th Infantry Division of Third Army, wa advancing with his platoon when they were suddenly raked by con cealed machine guns. Murphy felt a sharp pain in his shoulder, a though a white-hot blade had been thrust into his flesh to the bone.

Murphy's comrades scrambled for cover, but several of them had been hit and were calling for the medic. Ignoring his own painful wound, Murphy began slithering over the ground with his medical kit, tending to each injured man. Crawling toward yet another victim, the aidman detonated a mine, the explosion tossing him into the air. The blast had blown off one of his feet.

Dazed, frightened, and in deep pain, Murphy tried to bind up his bloody stump, a task made even more difficult by the bullets zipping around him and by his own shoulder wound. He heard the anguished cry of "Medic!" and began dragging himself toward another wounded man. There was a loud explosion. Murphy had set off a second mine. This one killed him.

Assaulted from three directions, the German First and Seventh Armies began to scramble back toward the Rhine. So clogged were the roads being used by the Wehrmacht that American fighter-bombers swooping down on German columns with bombs and machine guns could almost be assured a hit. American artillery, massed in large numbers, pounded the Feldgrau every step of the way, turning the withdrawal into a rout.

Most of the fleeing units ran out of gasoline, and the roads became blocked with abandoned, damaged, or wrecked vehicles, guns, and equipment, along with thousands of German bodies.

General Bob Frederick, leader of the 45th Infantry Division, was awed by what he saw in his zone of advance as a result of this enormous blasting by air and artillery. Talking to a Seventh Army staff officer about the situation confronting his advancing Thunderbirds, Frederick stated, "It is difficult to describe the destruction. Scarcely a man-made thing exists in our wake. It is even difficult to find buildings suitable for CPs." He added, "This is truly the scorched earth."

Zweibrücken, an industrial town in the path of General Frederick's advance, was now little more than rubble. Fires burned uncontrolled; neither water, firefighting apparatus, nor men were available to douse them. Most of Zweibrücken's 37,000 inhabitants had fled, leaving only a few hundred in cellars and caves.

In the Third Army zone, Combat Command B of Gaffey's 4th Armored Division roared off at 6:00 A.M. on March 20, brushed aside a force of 100 Germans and four self-propelled guns in the village of Walheim, and at 5:20 P.M. sent its first tanks into Worms, the Rhine city that had at first been the objective of Patch's Seventh Army. Elements of the 11th Armored, 90th Infantry, and 5th Infantry divisions advanced steadily during the next two days and cleared the west bank of the Rhine in the Mainz-Oppenheim-Worms sector.

Jubilation reigned at Patton's Third Army headquarters. "This is the most brilliant Patton campaign of them all," wrote Major General Hobart "Hap" Gay, Patton's chief of staff, in his diary.

Another Third Army staff officer summed up the sentiment there: "The goddamned Krauts are getting cross-eyed trying to watch the whole 500-mile stretch of the Rhine from Switzerland to Holland."[3]

The Saar-Palatinate campaign was over. Sandy Patch's Seventh Army and George Patton's Third had virtually wiped out the German First and Seventh Armies—precisely as Wehrmacht battle commanders had feared when they had asked to be pulled back over the Rhine in orderly fashion a week before.

"Tremendous losses in men and matériel," wrote the chief of staff of the German Seventh Army in his daily log.

Patch's army and attached French units had captured 22,000 prisoners in the Saar-Palatinate fight, while Patton's men inflicted 113,000 casualties on the Germans, if prisoners were included. Third Army casualties in the campaign numbered 5,220, while Patch's army, up against heavily fortified positions, suffered about 12,000 casualties.

Taking no chances of a repeat of the Ludendorff Bridge catastrophe, by March 19 the Germans had blown all the Rhine bridges from Ludwigshafen northward. The Maximilian Bridge was destroyed on March 21—by the Americans, when an artillery shell struck a detonator, setting off prepared dynamite charges. Another bridge at Speyer would go up on March 23, leaving only the bridge at Germersheim. During the night of March 22, German vehicles and artillery pieces that had survived the debacle in the Saar-Palatinate slipped over the Rhine on the Germersheim span.

The following day, March 23, Field Marshal Kesselring at Ziegenberg received a signal from Hitler: "Withdrawal of First and Seventh Armies over the Rhine is authorized."

There were few German survivors west of the river to withdraw.

George Patton wasted little time savoring his victory, hailed in American newspapers as one of history's most dazzling military operations. For Patton, the dash through the Saar and destruction of two German armies was merely the preliminary bout. Third Army was still on the "wrong" side of the Rhine—and Montgomery would be crossing the river in the north in 24 hours.

An aide handed Patton a telegram from General Leonard Gerow, a long-time friend and commander of the new Fifteenth Army, which had not yet been committed. "Congratulations," the signal blared, "on surrounding three armies—one of them American."[4] Gerow was referring to Patton's spearhead slicing in behind Sandy Patch's Seventh Army.

Since the morning of March 21, General Red Irwin had known that his 5th Infantry Division had been tapped to make a "sneak" crossing of the Rhine. "You've got to get across, Red," General Eddy told Irwin. "George has been cussing and raising hell for days." Irwin's men would sneak over at Oppenheim, not far from where Napoleon had crossed in 1806—a poetic touch appealing to George Patton.

On the following night, Matt Eddy arrived at Irwin's CP in an agitated state of mind. He told General Irwin that Patton had been "breathing down my neck all day" and "he insists on getting across tonight."

"Can't be done," Irwin responded. Neither all his units nor needed equipment had arrived yet. "Then you'll go tonight with what you've got," Eddy said solemnly.[5]

Before the 5th Infantry Division commander fell into exhausted slumber after midnight, he wrote in his personal diary:

Ample artillery is in position, and so far we have evidence of few enemy personnel or weapons on far shore on our front. . . . It is a big and difficult operation, however, and it is hard to predict the outcome at this time.

Eddy's corps headquarters had been working feverishly to organize an air-ferrying service to reinforce Irwin's assault troops on the far shore of the Rhine. The plan was unprecedented and carried out in great secrecy.

It was proposed that an entire battalion of infantrymen be taken across the Rhine, one soldier at a time, in a fleet of Cub aircraft, which came to be known as Maytag Messerschmitts. The operation was code-named, facetiously to a point, Grasshopper. Colonel Rod Gott and Major Tommy Haynes, both of XII Corps artillery headquarters, had assembled 91 of the "puddle jumpers" and had them ready to take to the air on the morning the 5th Infantry Division was to cross.

Colonel Gott and Major Haynes had tested the proposed operation the day before, ferrying one entire company and its equipment over a smaller river with such precision that they were convinced they could carry a battalion over the Rhine in less than three hours.

On the afternoon of March 21, two officers of the United States Navy, Lieutenant Commander William Leide and Lieutenant D. L. Spaulding, were driving the roads of the Saar, bound for General Matt Eddy's XII Corps headquarters. They felt a curious mixture of apprehension and excitement. The previous afternoon, Commander Leide, in charge of an amphibious force known simply as Naval Unit 2, had been informed at Third Army headquarters in Luxembourg that he and his men were to assist in a key operation to cross the Rhine at Oppenheim.

General Conklin, Patton's chief engineer, had told Leide that the naval unit was to be loaded and ready to move the following afternoon. "The road you are to use is still in German hands," Conklin advised Leide offhandedly. "But it will be captured by tomorrow morning."[6] Commander Leide resisted an urge to ask the general how he *knew* the road would be held by Americans then.

Leide's force had been attached to Third Army for five months and had been engaged largely in what the navy men considered makework projects—loading barbed wire onto railroad cars or painting 15,000 road signs. They were itching for some action.

Arriving at XII Corps headquarters at Worrstadt, which was 20 miles from Oppenheim, Leide and his XO (executive officer), Spaulding, learned that the crossing was to be a "sneak attack" and would take place that very night. They also were told that no reconnaissance of the crossing site had been made. Leide and Spaulding swallowed hard on being told jump-off would be only a few hours away—at 10:00 P.M. Their boats were still en route to Worrstadt.

Stillness hovered over the Rhine at Oppenheim that night. Having done all he could do to make the crossing a success, Red Irwin went to bed at 9:00 P.M., an hour before the jump-off. He could not sleep. Finally he switched on the light and, as was his custom to relieve tension, began reading a mystery novel.

At 9:45 P.M. men of two assault companies of the 5th Infantry Division began walking stealthily toward the Rhine 700 yards north of Oppenheim. All that could be heard was the muffled rustling of the men's gear and heavy breathing from those burdened with machine guns, boxes of ammunition, light mortars, and bazooka rockets.

"It's too goddamned quiet," a rifleman whispered to a comrade. "I don't like it. It's too goddamned quiet."

"Yeah," was the soft reply. "Those Heinies are up to something."

After covering the 300 yards to the river, the assault infantrymen could detect scores of other shadowy figures moving about. These were men of the 204th Engineer Combat Battalion, who had carried heavy assault boats to the bank and would be responsible for getting the infantrymen over the river. The engineers already had the craft lined up in the water. A gentle lapping of the current against the waiting boats sounded far too loud to the nervous men. Someone dropped a paddle.

"Be quiet, you son of a bitch," came a stage whisper from the darkness.

Lieutenant Irven Jacobs, leader of K Company, scrambled silently into a boat. Up and down the bank, assault teams were quietly getting into other craft and shoving off. The splash of the oars seemed to carry for miles on the water—most certainly to the alert Germans only 800 feet away on the far shore.

Lieutenant Jacobs, in the lead boat, looked cautiously over the front ramp. He could detect the dim outline of the bank only 100 feet away. Jacobs felt his heart pounding and was afraid its sound would alert the enemy. Along with each man in the assault, the lieutenant expected to be raked with machine-gun fire at any moment. But not a shot was fired. Instead, he felt the gentle crunch of his craft on the shore.

K Company men scrambled out of their boats and moved inland. Moments later seven startled Germans emerged from the darkness with their hands in the air. They were told to paddle themselves back to the other side—unescorted.

At the same time that Lieutenant Jacobs and his men were crossing, another company shoved off in Oppenheim and at midstream were shot at by German machine-gunners. The men kept paddling and reached the far shore without casualties, but now the enemy along the assault sector had been alerted, and several men in succeeding waves were wounded by small-arms fire.

Several hundred yards south of Oppenheim, another assault company led by Lieutenant William Randle met with heavy resistance after touching down on the far shore. Randle and his men wiped out a few German machine guns before pushing on inland, but Private First Class Paul Conn, Jr., a rifleman, was ordered to remain and pin down a troublesome enemy machine-gun crew which had to be bypassed. It would be an all-night affair for Conn, who had to scoop out a hole with his bare hands. Only 30 yards away, the German machine-gunners opened fire at anything that moved.

Conn remained in a cramped position until daylight. On occasion a jocular thought flashed through his mind during his lonely vigil: "Hell, no one knows I'm here. I'll jump in the Rhine and swim back to the other side!"[7] At dawn Conn peeked over the rim of his shallow trench and spotted 10 Germans manning the automatic weapon. They apparently were unaware of the American's presence nearby. Conn took aim with his Garand rifle and squeezed the trigger. The weapon was jammed.

Out of habit, the young rifleman glanced around for assistance, but there were no Americans in sight; they had all pushed inland. In desperation, he pulled the pins halfway from two grenades and charged the enemy machine-gun nest. Just as he was ready to toss the first grenade—he had pulled the pin all the way out—the German gunners spotted Conn running toward them and threw up their hands in surrender. The last remaining resistance along the Rhine's east bank had been eliminated.

Patton's sneak crossing had indeed taken the Germans totally by surprise. It was not until 30 minutes after midnight—two and a half hours after the assault began—that an occasional enemy artillery round began to fall on both banks around Oppenheim. Meanwhile, American reinforcements were pouring over. Armor of the 737th Tank Battalion, tracked vehicles of the 803rd Tank Destroyer Battalion, and amphibious tanks of the 748th Tank Battalion crossed on motorized barges.

In the meantime, Lieutenant Commander Leide, the leader of Naval Unit 2, was pacing the west bank of the Rhine. His DUKWs (amphibious trucks; "ducks" to the GIs) and LCVPs, which carried men or vehicles, had arrived and were in the water. But the assault had been going so smoothly that there was no immediate demand for the Navy's boats. Leide's executive officer, Lieutenant Spaulding, had reached the far shore at 3:05 A.M. to direct his unit's operations there, but he and Leide found that the GIs were too busy paddling across to give the Navy officers the passenger traffic they were soliciting.

But by 7:00 A.M. Leide was going full swing, and in the hours ahead the LCVPs and DUKWs would bring more than 15,000 foot soldiers and 71 tanks to the far shore, while bringing back countless American wounded and German prisoners on their return trips. After five months of frustrating make-work for Patton's Third Army, Navy Unit 2 had gone into business in a big way.

Some distance back from the Rhine on the west side, Colonel Rod Gott and Major Tommy Haynes were experiencing disappointment in their plan to air-ferry the reinforcing battalion of 5th Infantry Division soldiers across the Rhine in the fleet of Maytag Messerschmitts they had scrounged. So successful had been the crossing, and so far inland had Irwin's men gone, that Operation Grasshopper was cancelled.

At Namur, Belgium, that morning of March 23, General Omar Bradley was eating breakfast when he received a telephone call from

George Patton. "Brad, don't tell anyone, but I'm across!" he exclaimed, his voice tinged with excitement.

Bradley was silent a few moments. "Well, I'll be damned," he blurted finally. "You mean across the *Rhine?*"

"Sure I am. I sneaked a division over last night. But there are so few goddamned Krauts around they don't know it yet. So don't make any announcement. We'll keep it a secret until we see how it goes."[8]

Gaffey's 4th Armored Division would follow on the heels of Irwin's crossing, and Bradley authorized Patton to bring 10 divisions over the Rhine at Oppenheim.

Shortly after dinner that night, George Patton was again on the phone to Bradley. His foothold was firm on the far side. "Brad, for God's sake tell the world we're across," shouted the triumphant Patton into the telephone. "I want the world to know Third Army made it before Monty starts across!"[9]

That afternoon, Bradley had received orders from General Eisenhower to turn the limelight on American accomplishments—orders he was only too happy to comply with. The 12th Army Group commander held a press conference. Field Marshal Montgomery's widely trumpeted Plunder was to jump off the following day, and Bradley was determined to steal some of Monty's glory.

Needling Montgomery's grandiose buildup for a Rhine crossing, Bradley claimed that with Hodges and Patton on the far shore, the Americans could cross the Rhine at practically any point—"without air-bombing, airborne troops, or artillery bombardment."

While Bradley was getting in his verbal haymakers at Montgomery, east of the Rhine Kesselring was receiving an old friend at his headquarters at Ziegenberg. Obergruppenführer (SS General) Karl Wolff, a shadowy figure in the Nazi hierarchy who bore the vague title Plenipotentiary for the Wehrmacht in the rear of the Italian front, did not call to reminisce. He was at Ziegenberg on business—the business of treason.

For several weeks, General Wolff had been conducting clandestine negotiations with Allen Dulles, head of the Office of Strategic Services, America's fledgling intelligence and covert action agency.

Working through a Major Waibel of the Swiss Army, Wolff had been risking torture and execution to seek acceptable terms for the surrender of all Wehrmacht forces in Italy.

Kesselring had already been aware of Wolff's negotiations with OSS chief Dulles in Berne, Switzerland, but he was not prepared for the proposal the SS general put before him: that Kesselring surrender all German forces along the Rhine at the same time the Wehrmacht was capitulating in Italy. Despite Kesselring's knowledge that Patton had bolted the Rhine during the night, that Hodges had been on the east side for more than two weeks, and that Montgomery was preparing a mammoth blow in the north, the commander in chief west refused to participate in the Wolff plot.

"I will defend the soil of the Fatherland and continue to the end even if I die in the fighting," Kesselring told Wolff. He said that he personally owed everything to Hitler—his rank, his command, his decorations.[10]

In a more practical vein, the field marshal pointed out that two well-armed SS divisions were behind the front, and he was certain they would take action against him should he try to join in a mass surrender conspiracy.

Wolff shook hands with the field marshal and flew off to Italy. Kesselring went back to trying to plug the leaking German dike along the Rhine River, pondering the portent of Wolff's visit.

17

A Daring Plan
to Seize Berlin

As German armies along the Rhine were bracing for the Anglo-
Americans' next blow, a young Dutch citizen, Arie Bestebreurtje,
was seated in a battered old automobile parked along a dark, deserted
back road outside Namur, Belgium. Bestebreurtje was an Allied
spy—one of the best. He had been instructed to rendezvous with
unknown parties at midnight at this location to participate in an ultra-
secret operation behind enemy lines. His contact officer in American
intelligence, whom Bestebreurtje had never seen face to face, had told
him only that "it will be one of the most significant operations of the
war."

That was all that the blond Dutchman needed to know. He thrived
on danger and intrigue, and he hated the Nazis with the passion of a
man who had seen his nation under the yoke of an oppressive enemy
for five years. This mission, he believed, would provide another opportunity to strike a blow at what he called "the Hitler gang."

Bestebreurtje was a captain in the Dutch army, and a special agent
for the American OSS. A native of Nijmegen, where a bloody battle
had been fought the previous fall, he had already parachuted behind

German lines several times. The OSS considered him to have special knowledge of Berlin, the Nazi capital. He had lived in Berlin as a teenager, and knew the region intimately. During the war, he had slipped into Berlin from Switzerland and spent long periods sending back reports on conditions there, operating almost within the shadow of Hitler's Reich Chancellery.

Arie Bestebreurtje's name had proven a tongue-tangler to the Americans, so they had taken to calling him simply Captain Harry. He had parachuted into Holland with the U.S. 82nd Airborne Division the previous September 17 in the mammoth airborne assault code-named Operation Market-Garden. There he had gained a reputation among American airborne men as a courageous and exceptional officer whose forays behind German lines had provided valuable information.

Now seated alone along the dark road near Namur, Captain Harry pondered the nature of his mission. He had been told to remove all insignia from his uniform, even though the rendezvous was taking place deep in Allied-held territory. These Americans, Bestebreurtje reflected—they tend to get overdramatic sometimes in this spy business.

Down the road in the darkness he spotted the dimmed lights of a vehicle moving in his direction. This would be the American party he was to meet. Or could it be enemy agents?

The other car drove up alongside Captain Harry's vehicle and shut off its lights. Two shadowy figures emerged and walked up to the OSS agent. He could see that the strangers were dressed in American army officer's uniforms. Both men were strangers to him. The pair gave a prearranged password, and Bestebreurtje responded with the correct countersign. Hands were shaken, but few words spoken. The three men climbed into the American staff car and set out through the night.

Captain Harry knew the countryside well and could tell that the car was heading westward across Belgium, and he knew when the frontier was crossed into France. The journey continued through the night, and at daybreak the car drove into Sissone, a town a short distance east of Paris. Bestebreurtje still had no inkling of his mission.

He was taken to an American troop area outside the city, and from the insignia on soldiers there concluded that he was among members of the U.S. 82nd Airborne Division.

Captain Harry was escorted to the office of the commander, 38-year-old Major General James M. Gavin, who greeted the Dutch officer warmly. Gavin exchanged brief small talk with his old friend and comrade-in-arms, then told him, "Come with me, I want to show you something."[1]

General Gavin led Captain Harry to the closely guarded war room. The security precautions in effect were even stricter than normal. They passed through a second door to a room where a large curtain covered a wall map so that it could not be seen through the open door between the two rooms. Bestebreurtje had long been accustomed to this, but even he was impressed by such intense secrecy.

Gavin hauled the curtain off the map, and Captain Harry felt his heart skip a beat. He quickly recognized the locale—Berlin, the nerve center of Nazi Germany.

"This is it," the general observed with a trace of excitement in his voice. "Operation Eclipse—a parachute assault to seize Berlin."

Captain Harry said nothing for several seconds. He was frankly astonished by the audacity of the proposed airborne mission. Capturing hostile Berlin with a lightly armed force, some 325 miles *behind* German lines?

Gavin, known as "Slim Jim" to his men, quickly detailed Eclipse to the still nearly speechless but excited Bestebreurtje. The general and the handful of his staff and commanders who were in on the ultra-secret mission were sobered by the prospect, but at the same time the thought of parachuting into Berlin *was* exhilarating. Bestebreurtje not only was to provide the American airborne force with his intimate knowledge of the terrain and buildings around Berlin airports marked for seizure, but would jump with Gavin and his men into the Nazi capital.[2]

Allied airborne commanders were convinced that the time was right for Eclipse. In "normal" conditions, dropping a large force of airborne men into a teeming and heavily defended enemy metropolis 325 miles behind the lines would have resulted in the quick destruction of the invaders. But with the Rhine already breached and Field Marshal Montgomery's 21st Army Group poised to vault over the river, German forces were in disarray and demoralized.

Berlin itself had been almost totally denuded of troops, and its citizens were living in daily terror of the fate awaiting them when the Russians entered the capital. The Red Army by mid-March was only 50 miles away, near the Oder River. Even Hitler's bodyguard regiment, zealous SS men fanatically devoted to the Führer, had been sent to the eastern front. The oncoming Russians could be expected to give no quarter if they arrived in Berlin first. On the other hand, more humane treatment could be expected from Americans and the British. Odds were heavy that, after token resistance from Volkssturm units, Allied paratroopers might even be welcomed.

The primary advantage of Eclipse would be that the Western Allies could call the tune as to how the Russians were accommodated when they arrived in Berlin.[3]

Earlier, Major Barney Oldfield, a qualified paratrooper and chief press officer of General Simpson's Ninth Army, had been approached by a SHAEF colonel who, in utmost secrecy, confided plans for the mass parachute drop and glider landing in Berlin. He said that Eclipse would be launched "should Hitler be assassinated, or the Nazi government overthrown, or other eventuality arise which could be exploited."

The SHAEF officer said that a B-17 had been outfitted with a press-radio transmitter, and that two troop carrier planes were being allocated to carry a carefully selected group of war correspondents to Berlin. These reporters would be chosen from what the SHAEF officer called the Berlin List. Major Oldfield, the colonel stated, as a qualified paratrooper, would be escort officer for the correspondents. Presumably Oldfield would parachute into the Nazi capital with combat elements and be on hand to greet the writers on the Berlin List once a foothold had been secured. It was a sobering prospect. Yet, like other parachute officers, Oldfield wanted to be involved in an historic event of this significance.[4]

The Berlin airborne assault force was primed and ready to go. Its orders read: "Be prepared to enter Berlin, quell local disorders."

Those disorders, it was implied, could come from armed German soldiers, a chaotic civilian population—or Red Army troops bursting into the Nazi capital.

It was considered crucial that the invaders from the sky be a combined Allied force, but primarily American and British. Orders called for:

[1] Gavin's 82nd Airborne Division to seize airfields Tempelhof and Rangdorf, priority Tempelhof;

[2] Major General Maxwell Taylor's 101st Airborne Division to capture airfields Gatow and Staaken, priority Gatow, but be prepared to seize airfield Schönwald as an emergency alternate;

[3] A British brigade to capture airfield Oranienburg, with Schönwald as an emergency alternate;

[4] One regiment (possibly Polish) in reserve.

In his lengthy and detailed briefing of General Gavin and his key aides of the 82nd ("All American") Airborne (the outfit that the legendary Sergeant Alvin York first made famous in World War I, long before men conceived the notion of jumping out of airplanes and landing gliders en masse), Arie Bestebreurtje stressed that each building around Tempelhof was a miniature fortress. This district was known as the Mietskasernen. Based on Captain Harry's warning, Jim Gavin knew that, if the German army resisted, the Mietskasernen would be a tough nut to crack.[5]

Gavin had located an abandoned airfield a short distance from Sissone, and rehearsals for the Berlin operation were long since underway by the time Field Marshal Montgomery was nearly ready to bolt over the Rhine. General Taylor was holding similar "dry runs," as the troopers called them, with his 101st (Screaming Eagles) Airborne.

In mid-March, Lieutenant General Matthew Ridgway, the 48-year-old commander of the U.S. XVIII Airborne Corps, sent out a memorandum which stated the opinion that Eclipse would be a success and that the most important factor would be its psychological impact on the inhabitants of Berlin. "It is believed that the German mind will be tremendously influenced by the personal appearance and dress of individual soldiers who participate in Operation Eclipse," Ridgway wrote. "It is the intention of this headquarters to take into Eclipse a dress as well as a field uniform . . . and to insist on the highest standards of soldierly appearance."

As the German nation and its armed forces seemed to be on the verge of crumbling, with the Western Allies at the Rhine and the Russians at the Oder, the Anglo-American chambers of state were resounding with heated debate over Operation Eclipse. General Eisenhower, the supreme commander, in recent weeks had been acting more and more on his own in strategic matters without prior consultation with the Combined Chiefs of Staff. Opposing Eclipse, he argued that his was not a "political army" and that Berlin was not an important *military* target.

Winston Churchill, on the other hand, along with field marshals Brooke and Montgomery, were all for taking Berlin by any means possible and as soon as possible—ahead of the Russians. They pointed out that nothing would be gained if the Western Allies won the war strategically and lost it politically. "War is a political instrument," Churchill declared. "Once it's clear that you are going to win, political considerations must influence its further course."

Preparations and rehearsals for Eclipse were among the most closely guarded secrets of the war. Only those with an absolute need to know were privy to the details. Most officers and men in the airborne divisions had no inkling that an assault on Berlin from the sky was even being considered. To them, the training had merely been routine. Yet in spite of such intense secrecy, Eclipse became known to the Russian high command.[6]

Marshal Joseph Stalin, on learning of the impending plan of the Western Allies to seize Berlin in a massive airborne operation, called in the commanders of the two Russian army groups nearest to the Nazi capital, Marshals Georgi Zhukov and Ivan Koniev. To spur them on, Stalin told the two commanders that Berlin was Eisenhower's primary target, and that "two Allied airborne divisions are being rapidly readied for a drop on the German capital." He then pitted one marshal against the other, challenging each to get to Berlin first, ahead of the Americans and British.

With two American armies already over the Rhine, Field Marshal Montgomery was making final arrangements for Plunder, the vault by 21st Army Group. In the scope of its firepower, massed troops, airborne and naval support, Plunder would be second only to the D-Day landings in Normandy in its magnitude. In London, Fleet Street had pulled out all the stops: Plunder would be Montgomery's crowning achievement in the war, a fitting climax to the career of the man billed as Great Britain's finest soldier since Wellington and Kitchener.

For weeks an enormous wealth of supplies had been flowing to Montgomery's forces from Antwerp by freight car and by truck over the battered roads of Belgium, Holland, and Germany. As time for launching the Rhine assault neared, each of Montgomery's three field armies was receiving 10,000 tons of supplies per day. The flatlands of Holland and Belgium, stretching from Nijmegen all the way back to the Channel coast at Antwerp, was reminiscent of the English countryside prior to D-Day, packed with troops, ammunition and fuel dumps, guns, tanks, trucks, amphibious vehicles, assault craft, boats, pontoons, and bridging equipment.

To many, it appeared that Montgomery was preparing to launch an operation which would carry his armies to the ends of the earth. He left nothing to chance. Material was secured to warm the motors of assault craft in case the weather turned cold. In the event of extensive German flooding, he had stockpiled a half-million gallons of fog oil to be burned in 100,000 smoke generators. A half-million air photographs and 800,000 maps were distributed to all levels of command. Platoons and even squads studied the terrain they were to cross on the other side of the Rhine. Each night, patrols from 21st Army Group slipped over the water barrier along a 75-mile sector and brought back current information on German defensive positions.

Between March 20 and 22, warplanes of the U.S. Eighth and Ninth air forces and the Royal Air Force Bomber Command flew 16,000 sorties over the region in front of Montgomery's poised armies arrayed along the Rhine. They plastered enemy installations, rail centers, road junctions, bridges, and rolling stock with 49,500 tons of bombs, including the 22,000-pound monsters called "Grand Slams."

The Allied air barons referred to the saturation bombing as "isolating the battlefield."

The massing of nearly 1 million British, American, and Canadian troops—including the 80,000 to be in the assault—could not be concealed from German observers on the east bank. But beginning at dawn on March 21, a smoke screen was laid, which masked the specific movements of Montgomery's forces on the west bank. This screen, 75 miles long, would continue until H-hour for Plunder, some 60 hours later.

Bernard Montgomery, a master of the set-piece battle, would have under his command 17 infantry, 8 armored, and 2 airborne divisions, 13 of which were American, 12 British, and 2 Canadian. Added to these were five armored brigades, a British commando brigade, and a Canadian infantry brigade.

Supporting the amphibious crossing would be a mass parachute drop and glider landing by the XVIII Airborne Corps under General Ridgway. His force would consist of the U.S. 17th Airborne Division, which would be making its first parachute and glider landing but had seen heavy action in previous months during the Battle of the Bulge; and the British 6th Airborne Division, veterans of the D-Day drop in Normandy. Code-named Operation Varsity, the airborne assault would be a monumental operation in its own right. More parachute and glider troops would land in one day than had ever been attempted.

As D-day for Plunder drew closer, Major Barney Oldfield was up to his ears in efforts to accommodate the swarm of correspondents who had descended on 21st Army Group to cover the "first assault crossing of the Rhine," as it was being billed. The even-tempered Simpson had been working in harmony with Montgomery. But Major Oldfield and other staff officers at Ninth Army privately felt that 21st Army Group headquarters, encouraged by the field marshal himself, was seeking to ensure that Montgomery was the one in the spotlight when Plunder was launched, even though nearly half of the troops involved would be American.

A simmering controversy arose between American and British correspondents in 21st Army Group, and increased in intensity as the March 23 H-hour drew closer. American reporters were concerned that the achievements of the unassuming Bill Simpson and his Ninth Army would not fare well under British press censorship, and that Fleet Street writers would advertise Plunder as a "British" or "Allied" operation.

The suspicious American war correspondents felt justified when Major Oldfield was instructed that 21st Army Group could only be called "an Allied force." American leadership was not to be mentioned, at least at the outset, and none of the units taking part in Plunder/Varsity could be identified until specific approval had been given by Lieutenant Colonel Pat Saunders, Montgomery's chief censor.

In this atmosphere of journalistic jealousy, American and British writers assembled at Schloss Rheydt on March 21 for a briefing by General Simpson on the forthcoming assault over the Rhine. The Ninth Army commander went into detail on his 16-division force, now at its greatest strength, whose 30th and 79th Infantry Divisions, both veteran outfits, would spearhead the assault.

Completing his briefing, Simpson asked if there were any questions. The first one was by Wes Gallagher, an American reporter for the Associated Press. "How soon will Ninth Army be released for publication in this operation?" he inquired. The British writers eyed Gallagher suspiciously.

"I don't know," Simpson responded in his customary candor. "I think Field Marshal Montgomery will announce us in due time."

If General Simpson recognized the dissatisfaction implicit in Gallagher's loaded question, he gave no indication of it.

Major Oldfield and other Ninth Army press officers were notified by 21st Army Group a few days before Plunder that tough censorship restrictions would be scrupulously enforced. H-hour would be about 9:00 P.M. on March 23, but the new orders prohibited any mention of the Rhine crossing until 4:00 P.M. the following day. American correspondents were furious. They suspected Bernard Montgomery machinations were involved, as the 4:00 P.M. release time would assure a clean sweep for British newspapers, the bulk of which were morning editions. This strict time peg and the clamp on mentioning the U.S. Ninth Army made it certain that Montgomery would gain full credit for Plunder, the American scribes fumed.

"I can see the London headlines now," an American wire-service correspondent mused. "'Monty Storms Over Rhine.'"

For their part, British war correspondents felt that most of the complaints by their American opposite numbers were without foundation. "Why shouldn't London newspapers focus on British activities in the war?" they asked. Newspapers throughout the world have traditionally highlighted "local" events which are of paramount interest to their readers, British newsmen pointed out.

As for Montgomery's edict that the 21st Army Group would be referred to only as "an Allied force," the British reporters declared that the field marshal only was stressing Allied unity, and at this stage of the war it would be disruptive for media to begin pointing out which nation's armed forces had achieved what in any particular action.

Just as American correspondents felt that Montgomery was behind much of the friction between the Allies, the British reporters held a similar view toward the man they believed to be the chief culprit on the American side—the flamboyant General George Smith Patton. Patton, the British held, was arrogant, rude, egotistical, and addicted to lashing out needlessly at the British in general and Bernard Montgomery in particular.

On the night of March 22, the gears of the gigantic Plunder machine began to turn. In Simpson's Ninth Army zone, grim-faced assault troops of Leland Hobbs's 30th (Old Hickory) Division and Ira Wyche's 79th (Cross of Lorraine) Division began trudging toward jump-off positions. Shoulder patches had been removed and all insignia on vehicles painted over. The enormous smokescreen that had blanketed much of the Rhine for hours had become a source of irritation to some of the assault troops.

At the same time these American assault elements were moving into position, a special group of troops with shoulder patches and vehicle markings of the 30th and 79th divisions left the same area and headed southward along the Rhine. These specialists carried a large number of radios in their vehicles over which they sent out a constant stream of phony messages, which conveyed the impression to German monitors that an entire Ninth Army corps was going to assault the Rhine far to the south.

Along the east bank of the Rhine, a German army already mortally battered in the previous weeks' fighting west of the river dug in and waited grimly for Field Marshal Montgomery's enormous blow to strike. "This is a shadow army," the chief of staff of Army Group G wrote in his daily log. He added that the German officer corps "wondered just what were the demands of duty."

Shadow army or not, there remained a hard core of Feldgrau on the east bank of the Rhine who would fight to the end when the Anglo-Americans stormed the river in Montgomery's sector. Many still had faith in Hitler, and respect for his stand-or-die edict.

General Johannes Blaskowitz, commander of Army Group H, estimated that Montgomery's assault would strike between Emmerich and Dinslaken, the latter seven miles south of Wesel. It was at Wesel, Blaskowitz rightly guessed, that the Anglo-Americans would make their major effort. Wesel, with a peacetime population of 24,000, was the hub of a rail and road network. This crucial sector would be defended by the First Parachute Army, now whittled down to corps size, commanded by General Alfred Schlemm, who had given the Americans fits in recent days on the west bank of the Rhine as they sought to seize the Wesel bridges.

German forces facing Montgomery on the east side of the Rhine had only hasty field fortifications along the bank and a railroad embankment that paralleled the river, but had a fairly solid line to oppose the crossing, backed by a moderate amount of firepower. Fearing parachute and glider landings, Blaskowitz had virtually stripped his air defenses in Holland and brought 81 heavy and 252 light anti-aircraft guns to the vicinity of Wesel, where he expected the airborne assault to strike.

On the morning of March 21, General Schlemm was at his desk at headquarters of the First Parachute Army. He was aware Montgomery was about ready to jump off and was making last-minute efforts to bulwark his defenses. Suddenly, he heard a shout: "*Jabos! Jabos!*

(fighter bombers!)" There was a scramble for cover as the whine of diving aircraft echoed over the terrain, followed by the frightening sounds of bombs whistling through space. Then, a terrific explosion. Thick black smoke filled the headquarters house. Wood and glass splinters flew; General Schlemm was pulled out from under debris, bleeding profusely from numerous wounds.

Soon running a high fever and bedridden with painful injuries, Schlemm refused to be evacuated. He continued to direct preparations to meet Montgomery's assault, but was in no condition to exert any real influence on the battle ahead. General Schlemm's serious wounding was a heavy blow to the declining morale of the Wehrmacht along the Rhine.

At midafternoon on March 23, the British field marshal was in consultation with his commander and staff. His armies were ready to attack. A last-minute report from his meteorologists was analyzed to determine if conditions were right for the airborne phase of the assault. They were. At 3:32 P.M. Montgomery gave the order: *"Go!"* Officers scurried from the meeting to flash the code word to hundreds of waiting units: "Two if by Sea." The British were coming.

Montgomery's exhortation to the troops had been printed ahead of time. Now thousands of them were hurriedly distributed:

The enemy possibly thinks he is safe behind this great river obstacle. We all agree that it is a great obstacle; but we will show the enemy that he is far from safe behind it. This great Allied fighting machine, composed of integrated land and air forces, will deal with the problem in no uncertain manner.

And having crossed the Rhine, we will crack about in the plains of Northern Germany, chasing the enemy from pillar to post.

Over the Rhine, then, let us go. And good hunting to you all on the other side.

May "The Lord mighty in battle" give us the victory in this our latest undertaking, as He has done in all our battles since we landed in Normandy on D-Day.

It was shortly after 3:00 P.M. that same day when Winston Churchill, his aide Commander C. R. "Tommy" Thompson, and secretary Jock

203

Colville climbed aboard an RAF plane at an airport in Middlesex, England, and flew to Venlo, Holland, near the German border. The Churchill party was driven to Bernard Montgomery's headquarters, where the field marshal greeted the prime minister in corduroy trousers and a pullover sweater frayed by long use. Tea was served. In deference to Montgomery's strict no-smoking, no-drinking edicts, Churchill was careful to be outside when he lit his long black cigars and waited until the field marshal had gone to bed before availing himself of some Scotch that had mysteriously been smuggled in.

For days, Churchill had been a source of worry to Field Marshal Brooke. The prime minister was determined to witness the launching of Plunder and even spoke of "going up to the river in a tank." Knowing that it was futile to continue to protest, Brooke did the next best thing: he asked Montgomery to keep Churchill at 21st Army Group headquarters at Venlo, far behind the lines. The prime minister had other ideas. Not only did he intend to watch the Rhine attack, but he planned on crossing to the other side of the moat on the heels of assault troops.

At Bletchley Park north of London, Ultra had detected an atmosphere of panic among besieged German commanders at all levels. A series of signals had been intercepted and decoded at Station X which warned of a looming Allied airborne assault on the east bank of the Rhine between Wesel and Duisburg.

Station X technicians and evaluating officers, on the basis of Ultra intercepts of German top-secret signals, were gaining a picture of an Adolf Hitler hallucinating in the confines of his Reich Chancellery bomb shelter. The Führer sent a signal to Kesselring on March 20 with the good news that hard-pressed German armies in the west could expect a shipment of 1,331 tanks, armored cars, and assault guns from the Third Reich's March industrial production. These armored vehicles never arrived; they existed only in Hitler's mind.[7]

Three days later, German armaments chief Albert Speer sat in on a conference with Hitler and was astonished to hear him discussing nonexistent steel production. Speer was even more amazed to hear the Führer wax enthusiastic over five new tank models which would be demonstrated soon. These tanks didn't exist, either.[8]

18

"Two If by Sea"

Late in the afternoon of March 23, Major General William M. "Bud" Miley, a former West Point gymnast who led the U. S. 17th Airborne Division, and Major General Eric Bols, commander of the British 6th Airborne Division, received the coded signal: "Two if by sea." Operation Varsity, the parachute and glider assault over the Rhine, would strike in the morning.

German forces defending the river would be hit by an enormous thunderclap from the blue. Within a period of just over two and a half hours, 21,692 Allied paratroopers and glidermen would descend on the dug-in Wehrmacht, along with 614 jeeps, 286 guns and mortars, and hundreds of tons of ammunition, food, fuel, medical supplies, and other equipment. It would be the mightiest and most complex simultaneous airborne operation in history.

Never before had two entire airborne divisions been flown into battle in a single continuous effort. Larger numbers of fighting men had been delivered in both Operation Neptune (Normandy) and Operation Market (Holland), but those operations had been carried out over several days so that planes and crews could be used more than once.

The British and American forces in Varsity would touch down between Wesel and Emmerich in support of the British Second Army, which would storm the Rhine. American air forces would play the major role in flying the paratroopers and glider fighters into battle. The U.S. IX Troop Carrier Command, led by Major General Paul L. Williams, would transport paratroopers from both airborne divisions and glidermen of the 17th Airborne Division. Two groups of the Royal Air Force would tow the 6th Airborne gliders over the Rhine from England.

It had been just under five years since the world's first glider assault, on May 10, 1940, when 11 small German gliders carrying seven or eight men each had landed on a Belgian fort, Eben Emael, during Hitler's overrunning of the Low Countries. Now, the Anglo-Americans would launch 1,326 gliders—some of them large enough to carry bulldozers, trucks, and howitzers—in a stream some 150 miles long.

To increase the shock effect of Varsity, for the first time extensive use would be made of double-tow techniques; American C-47s would pull two Waco gliders, instead of only one as had been the previous practice. This was something of a gamble on the part of the Allied high command; an attempt to use double-tows for an airborne mission in Burma in 1944 had resulted in near disaster.

Allied airborne planners were having sleepless nights over the awkward setup for Varsity headquarters, which would be located in three widely separate places: Paris, Brussels, and Mark's Hall in southern England. Routine control and supervision of Varsity would be exercised partly from Maison Lafitte, headquarters of the First Airborne Army outside Paris, and partly from Mark's Hall, headquarters of the British 38th Carrier Group. British Air Marshal Sir Arthur Coningham was in command of all operating air forces, and his headquarters was in Brussels. Since Coningham and Lieutenant General Lewis H. Brereton, commander of First Airborne Army, might have to make joint decisions during Varsity, it was decided that Brereton would go to Coningham's headquarters at Brussels just before the operation was launched.

At Station X in Bletchley Park, Ultra had identified German units east of the Rhine in the region of the crossing. Ten Wehrmacht divisions had congregated where the Germans expected Field Marshal

Montgomery to hit. Several of these were nominally first-rate para-troopers and panzer formations, but all divisions had been cut to pieces in the fierce fighting west of the Rhine. At full strength this 10-division force would have numbered some 140,000 men, but when Plunder and Varsity struck, only about 50,000 Germans would be available to ward off the massive blow. Some 11,000 enemy fighting men would be in the precise locale of the crossing, intercepts by Ultra revealed, the others being held at other points in the region to be rapidly dispatched to the scene of the fighting.

The 17th (Thunder from Heaven) Airborne would take off from 17 airfields in north-central France in a region bounded by Rheims, Orléans, Evreux, and Amiens. The Red Devils of the 6th Airborne would depart from 11 airfields in southeastern England. Flying this elite force into battle would be 1,572 transport planes and 1,326 glid-ers, escorted by 889 fighter planes.

On March 22, the Thunder from Heaven division in France and the Red Devils in England had been "sealed in" at camps surrounded by thick strands of barbed wire. Alert guards manned the few exits and patrolled the grounds with orders to take whatever action was necessary to keep the airborne men penned up. If word seeped out and reached enemy ears, Varsity could end in disaster.

For two weeks, Generals Miley and Bols had been having sleepless nights over mounting evidence that there had been serious security leaks. Their concern heightened when Günther Weber, a news com-mentator at Radio Berlin, trumpeted over the airwaves: "Allied air-borne landings on a large scale, to establish bridgeheads east of the Rhine, can be expected. We are prepared."

Speculation as to the locale and timing of the Allied airborne attack had been rampant in American newspapers and on radio. In the towns and villages around Châlons-sur-Marne, where Miley was headquartered, it had become common knowledge that the men with the golden eagle's claw shoulder patch were about ready to leap Hitler's final barricade. There was no way to conceal from spying eyes the presence of the hundreds of gliders and troop-carrier aircraft squatting on airfields in the region.

There was no doubt that the Wehrmacht high command knew that the operation was imminent. A German order had been captured outlining the defense measures dealing with the Anglo-Americans. The plan called for small groups of German troops, mainly parachute and SS soldiers, to be stationed at all locales east of the Rhine where landing from the air might be expected. Their job was to attack the invaders when they were the most vulnerable—when they were scrambling out of parachute harnesses or hacking their way out of gliders.

And if there had been any doubt among German commanders as to where the Allied airborne operation would strike, it was dispelled when Field Marshal Montgomery laid the 70-mile smoke screen stretching along the Rhine between Rees and Wesel.

Now, on the eve of D-day for Varsity, General Eric Bols was at his headquarters at Aldershot, England, methodically supervising the last-minute preparations of his battle-tested 6th Airborne Division. His Red Devils had helped spearhead the invasion of Hitler's continental Europe in Normandy the previous June, but this would be the first time Bols had led the division in a combat operation.

Ten days before, General Bols had held a final briefing for his top commanders. One officer present reflected that Bols detailed his division's role in the Rhine assault as calmly "as though he were outlining plans for a church social."

At that same conference, Brigadier James Hill, leader of the 3rd Parachute Brigade, had infected his fellow commanders with his enthusiasm and confidence. "Our artillery and air support is fantastic," Brigadier Hill exclaimed. "If anyone is worried about the reception you'll get, put yourself in the place of the enemy. Beaten and demoralized, pounded by our artillery and bombers, what would you think if you saw a horde of ferocious, bloodthirsty paratroopers bristling with weapons, cascading down upon you from the skies?"

The fire-eating Hill closed with a straight-faced observation: "But if, by chance, you should happen to meet one of these Huns in person, you will treat him with extreme disfavor." Gales of laughter erupted over his understatement.

Two days later, General Bols and his top commanders had flown to Châlons-sur-Marne, to meet with the leaders of the 17th Airborne Division and to attend a briefing by Miley and his regimental colonels. The peppery Miley detailed his division's plans for the air landing across the Rhine. Projecting great confidence, even though he and most of his men would be jumping into battle for the first time, General Miley punctuated his presentation with bits of humor.

General Bols and his battle commanders were uneasy when Miley concluded. The American plan seemed to them extremely hazy. Accustomed to briefings in much more detail, the British felt that the "newcomers to war" were going at their task in a haphazard manner. In fact, General Miley's plan was about as detailed as that of the British airborne; it was his style to try to conserve valuable time by not going deeply into explanation for his Allied counterparts.

As Miley took his seat, Colonel Edson D. Raff, leader of his 507th Parachute Infantry Regiment, got to his feet. British officers present knew Raff well—and they disliked him. The feeling was mutual. Raff, one of the American army's most experienced combat officers, had clashed often with his British superiors during the Allied invasion of North Africa in November of 1942. At that time Raff had led the 509th Parachute Infantry Battalion, which had spearheaded the operation. It was the first mass combat jump ever by American paratroopers. After the Allies were established ashore and bitter fighting erupted in Tunisia, high British commanders, under whom the 509th Parachute Infantry Battalion was fighting, split up Raff's elite force and fed parts of it into the line, often in platoon units, to fight as straight infantry.[1] Colonel Raff had been furious over what he considered a "foolish" deployment in little packets of his specialist airborne battalion, and he had minced no words in telling his British superiors of his feelings.

Now, in front of a group of British officers, Raff was called on by General Miley to "present your regiment's plan."

"Well, general, it's like this," the 34-year-old Raff began. "In these woods [across the Rhine] there's a bunch of Heinies. We drop here"—he pointed to a large map—"about a half mile away, and we just go and root 'em out. We fly in west to east and I jump Number One from the lead ship."

His 20-second presentation concluded, Colonel Raff sat down.

Raff had hardly enhanced his reputation among his British counterparts, although they respected him as a tough, resourceful, and experienced battle leader. For his part, Raff could not care less what the British thought of his candor—even when the officer he was addressing outranked him. In early 1943, after the savage fighting in Tunisia, a high-ranking British general in London had asked Raff for his view on Major General K. A. N. Anderson as a battle commander. (Anderson was the British leader who had split up Raff's parachute battalion).

"What do I think of General Anderson as a battle leader?" Raff had snapped. "I think you should ship him home and stick him behind a desk."

Now, on the evening of March 23, with the massive airborne assault over the Rhine but hours away, men of Miley's 17th Airborne Division, each in his own way, were steeling their spirits for the greatest ordeal of their young lives—their first combat jump and glider landing. They gathered in pairs and little knots. Talk was meager and subdued. Feeble efforts by some to joke were met with scattered snickers and dull stares. Somehow, humor had vanished from their lives. In its place was the specter of what awaited them on the far side of the Rhine in the morning.

Loudspeakers blared out "In the Mood," "String of Pearls," "Paper Doll," and other hit tunes of the day. But the musical morale booster found few responsive ears. As the minutes ticked by at the 17th Airborne encampments, tension, thick and suffocating, heightened. The men checked and rechecked their Tommy guns, rifles, BARs, and machine guns. Many scrawled last-minute letters to the home folks. Pocket bibles were read intently. Catholic chaplains were hearing confessions, and brief religious services were held for other denominations. The chaplains had no shortage of "customers." On the eve of what promised to be a savage battle and with unknown terrors lurking across the waters, many men barely conversant with spiritual matters were struck by what the troopers called "fits of religion."

At an airfield east of Châlons-sur-Marne, site of a bloody battle in World War I, Private First Class John F. Magill, Jr., a 21-year-old paratrooper with a forward-observer section, was trying to console a badly frightened replacement radio operator. Magill had seen heavy action in the Battle of the Bulge; the 18-year-old replacement was going into combat for the first time on the morrow.

Putting an arm around the boy three years his junior, Magill, of Sugarloaf, Pennsylvania, observed in a fatherly tone, "It's not so bad once you get used to it." He neglected to mention that few ever got "used to it."

Rumors were rampant. It was whispered that fanatical German troops on the drop and landing zones had taken personal blood oaths to Hitler that they would stand and die defending the airhead; old men, women, and children had been trained to maim or poison the airborne invaders; in his desperation, Hitler had ordered poison gas to be used; tens of thousands of needle-sharp sticks had been placed in the drop zones to skewer descending parachutists. The rumors were inexhaustable and concocted out of the ingenuity of fear.

Aside from these nebulous reports, Allied intelligence officers produced solid evidence that the Germans apparently knew where Varsity would strike. The enemy had moved in large numbers of antiaircraft guns around the drop and landing zones in the vicinity of Wesel. Aerial photos showed clearly the snouts of these menacing weapons, pointing skyward and ready to open fire on the lumbering, low-flying troop carrier planes and tug-and-glider combinations.

In his stifling tent, trooper Lou Varrone, a 25-year-old rifleman in Raff's 507th Parachute Infantry Regiment, was idly twisting the dial on his small radio. Suddenly he heard a sultry female voice, which he recognized instantly.

"Hey, guys," the New York City native called to his comrades, "I've got the Berlin Bitch on the radio!" There was a scramble to crowd around the wireless set.

The "Berlin Bitch" was the sobriquet given to a Radio Berlin propaganda personality who played popular American tunes interspersed with detailed and lurid descriptions of sexual activities. Third Reich Propaganda Minister Josef Goebbels reasoned that the surest way to attract a large audience of American fighting men was for his female propagandist to broadcast large doses of explicit verbal sex. To heighten her credibility with Allied fighting men on the western front, the Berlin Bitch was fed up-to-the-minute war information by German intelligence agencies.

As Varrone and his comrades listened intently, the Berlin Bitch cooed over the airways:

"Hi, all you sexy guys in the 507th and 513th Parachute Infantry Regiments there in France. The British are getting you ready to be slaughtered for the greater glory of the King and the Empire, aren't they? And how do you fellows in the 507th like being penned up in that old French prison compound? They could have found a better place for you to spend your last night on earth, couldn't they?"

Varrone and the others exchanged solemn glances. Indeed they were "penned up in an old French prison yard."

"We know you're coming at twelve-oh-one [12:01 P.M.] tomorrow. And we know exactly where you're coming—at Wesel. Ten crack divisions from the Russian front will be waiting as a reception committee. So don't worry about your landing. Flak will be so thick you can simply walk down from the sky."

The voice, speaking in flawless, unaccented English, paused for effect, then added: "The only sensible thing for you to do tomorrow, when and if you get across the Rhine, is to promptly surrender. We'll give you good food and treat you well."

Someone flipped off the radio. There was brief silence. Finally a voice called out, "Hell, the old bitch doesn't know so much. We're supposed to land two hours *before* she says we are." The grim-faced troopers filed out of the tent.

Across the English Channel, Bols's 6th Airborne Division had been trucked to isolated internment camps in East Anglia and had spent most of the four days there in briefings, writing letters home, and passing on the "latest authentic rumors." Mostly it was just waiting. Waiting. And more waiting.

Eric Bols, quietly determined on the eve of battle, was not unduly concerned that his Red Devils were not up to full strength. The 6th Airborne, under Major General Richard Gale, had been badly chewed up in Normandy the previous June, and Great Britain, an island of some 40 million people, had been engaged in total war for more than five years. Its manpower was nearing exhaustion. Bols was convinced his airborne warriors, despite being under strength, would give a good account of themselves.

Soon—far too soon—the sky in the west was aglow with a setting sun. Most of the Red Devils in England and the Thunder from

Heaven men in France gazed wistfully at the kaleidoscope of color. It would be the last sunset many of them would see, and all of them knew it.

Meanwhile that late afternoon of March 23, Lieutenant General Miles Dempsey, commander of the British Second Army, was sifting through last-minute intelligence reports and resolving final snags. The 49-year-old Dempsey had been entrusted by Field Marshal Montgomery with launching the combined water-and-airborne attack across the Rhine.

During his service as a corps commander in Sicily and Italy, and later as leader of Second Army in France, General Dempsey had gained a reputation for making a major attack only when he was convinced he had the manpower, firepower, and supply advantage that would make success a virtual certainty. In his methodical approach to war, Dempsy was very much in the mold of his boss, Montgomery.

Dempsey had been preparing to cross the Rhine for many weeks. Now all was in readiness. His army would ferry over the water that night on the heels of an enormous artillery barrage between Emmerich and Wesel. Dempsey had long been concerned about the Diersfordter Forest, which stretched for five miles along high ground three miles east of the Rhine in his zone of advance. Concealed in that thick forest, Dempsey feared, would be many batteries of German artillery, which would pound his assault troops as they were crossing the river.

Another terrain obstacle that had bothered the British army commander and his planners was the Issel River, some three miles farther back from the Diersfordter Forest, which ran parallel to the Rhine. Aerial photos had revealed that a few bridges still stood over the Issel, and if they could be seized intact, Dempsey's plunge inland would be much more rapid. The mission of Bols's 6th and Miley's 17th Airborne Divisions was to seize these Issel bridges and to rapidly clear the enemy from the Diersfordter Forest.

A few weeks previously, the methodical Miles Dempsey had startled other American and British commanders—particularly the airborne leaders—by reversing accepted tactical doctrine for the employment of parachute and glider troops. Instead of Allied airborne troops landing ahead of amphibious forces, as had been the case in

213

North Africa, Sicily, Normandy, and Holland, Dempsey's ground troops would cross the Rhine under cover of darkness, and Bols's and Miley's parachutists and glidermen would jump and land a few hours later, after daylight.

Luftwaffe night fighters were still highly active, and Dempsey feared that his airborne flights would be riddled by the German aircraft if he sent them in during darkness. The Second Army commander was also haunted by the destruction of a large portion of the British 1st Airborne Division in Holland the previous fall when ground forces could not link up with the Red Devils at Arnhem. This time Dempsey wanted to make certain the link-up would occur in rapid fashion.[2]

Both of the airborne divisions would be components of the U.S. XVIII Airborne Corps, commanded by the stern, no-nonsense Lieutenant General Matthew B. Ridgway. Miley's 17th Airborne would come down in the southern portion of the XVIII Corps' zone, and the Red Devils would land on the northern portion. Measuring only five miles deep and six miles wide, the airhead east of the Rhine would be the most congested concentration of parachute and glider troops yet attempted, with 21,692 airborne men vying for space. Not only would Miley's and Bols's men be coming down on the heads of German troops, they were in danger of landing on each other.

Across the Rhine River in the vicinity of Wesel on the evening of March 23, General of Parachute Troops Eugen Meindl, leader of the II Parachute Corps, was aware that his battered force was about to be struck by all the lethal fury that the Western Allies could gather. Meindl's corps consisted of remnants of the 6th, 7th, and 8th parachute divisions, each whittled down by weeks of heavy combat to 3,500 to 4,000 men.

In recent days General Meindl had received a draft of former Luftwaffe ground-service personnel who were handed rifles and told that they were now infantrymen. "They're more of a hindrance than a benefit," Meindl snorted to his staff in disgust.

Meindl, who had seen long combat service, was pessimistic about being able to halt the Allied thunderclap. There was a serious short-

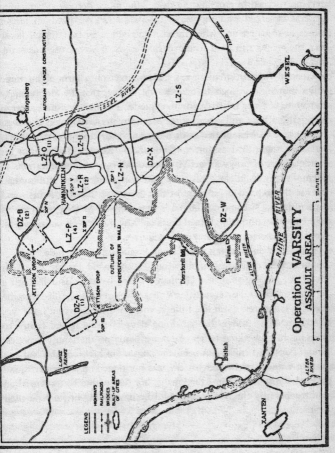

Operation Varsity Assault Area

age of weapons of all kinds, and artillery and mortar shells had to be rationed. Transport was in short supply, or nonexistent.

"With his huge quantities of ammunition and matériel, Montgomery can succeed in crossing the river," the much decorated parachute general observed philosophically to his staff. "We need no longer worry as to where and when the crossing will take place, for we know that the enemy can do it wherever he pleases, if he but uses his matériel in a proper way."[3]

General Meindl, at his corps headquarters on a farm about three miles southeast of Bocholt, had known for days that the Montgomery dynamite keg was nearly ready to explode. The massive smoke screen along the Rhine and observation of enormous activity on the west bank had told Meindl that. But now, on the afternoon of March 23, he knew the Allied assault was imminent. Reports flowed into his headquarters of heavy air activity against German communications to Meindl's rear. And along his river front for miles, Allied artillery had drastically stepped up its rate of fire. So heavy had the British and American fire become on the roads that General Meindl could maintain contact with his divisional units only by motorcycle couriers who traveled over back roads and cross-country.

"Well, it won't be long now," Meindl said softly to an aide.[4]

Some four miles east of the Rhine, 19-year-old Private Friedrich Baur was helping to man a German artillery battery. Baur had left a farm in the Black Forest nearly a year ago when he was called into the service. His parents still lived there, and Friedrich's fervent hope was that he would survive the remaining days of the war and return to help his hard-working father eke out a living on the family's modest lands. Neither artillery gunner Baur nor any of his comrades had any way of knowing that their battery was positioned on a plot of ground labeled "Drop Zone" on Allied maps, nor that within hours American paratroopers by the hundreds would be descending literally on their heads.

19

Crossing the Moat by Buffalo

As Operation Plunder was ready to strike the beleaguered Germans along the east bank of the Rhine the night of March 23, Adolf Hitler was holding one of his nightly conferences in the dank underground bunker at the Reich Chancellery in Berlin. The Führer did not appear to be unduly concerned about Montgomery's juggernaut. Those present knew all too well the reason for his apparent lack of alarm: Hitler was living in a fantasy world.

On this night he was issuing orders to his staff officers to bring up designated divisions or corps to help Field Marshal Kesselring ward off the Allied blow in the west. But most of these formations were either down to skeleton manning levels or nonexistent. None of those present dared inform Hitler of this fact.

In the upper levels of the German high command at this stage of the war there was little outward recognition of approaching catastrophe. Day-to-day activities went on much as they had in the beginning of the war. Hitler and his immediate staff were bogged down in a dream world of armies, corps, and divisions ready to fight. Often when orders flowed out of the *Oberkommando der Wehrmacht* in Berlin, there was no one in the field to receive them.

During recent days, Hitler had proposed withdrawing from the Geneva Convention as a "protest against the bombing of civilians" and to strengthen the nation's will to continue to resist. In this instance, he was talked out of the proposal. But to every Wehrmacht fighting man went a threat from Hitler himself: any German soldier taken prisoner unwounded would see his loved ones at home answer for his "cowardice and betrayal." Another order from the Führer directed that, before another foot of ground was relinquished to the enemy, all industrial establishments and supply depots in the Third Reich were to be blown up—regardless of the effect such actions might have on the German civilian population.

Now, with Allied troops crossing the Rhine in assault boats at that precise moment, the Adolf Hitler who in his glory days had dwelt on the fate of nations and the employment of his powerful legions was reduced to a discussion of trivia. Major General Fritz Burgdorf, Hitler's chief military aide, told him, "Minister Goebbels is seeking permission to turn the east-west avenue [in Berlin] into a runway. It would be necessary to remove the lampposts and clear the Tiergarten for twenty meters on either side."[1]

"Yes, he can do that," Hitler responded. "But I don't think he has to do all that. Fifty meters is wide enough."

A discussion followed as to whether the ground along the proposed runway in pulverized Berlin should be cleared to a width of 40 or 50 meters.

A short review of a small unit trapped on the eastern front ensued, after which General Burgdorf returned to the question of the Berlin runway. "Can I give Minister Goebbels approval to proceed?"

"Yes," Hitler said, "but I don't see why it has to be made so wide."

Colonel Nickolaus von Below broke in: "If JU-52s are going to land in the dark, the lampposts will cause trouble."

"All right, the lampposts can go," the Führer ruled. "But there's no need to level twenty to thirty meters of the Tiergarten on either side."

"I don't believe leveling for twenty meters is necessary," von Below declared, "but we must get rid of the lampposts."

"He can remove the lampposts," Hitler stated.

"Then I can tell Herr Goebbels to proceed?" General Burgdorf asked once more.

"Yes, he can proceed."

The question of a few Berlin lampposts had been resolved at the highest level.

At 21st Army Group headquarters at Venlo, Holland, that night, Field Marshal Montgomery took leave of his guests, Winston Churchill and Alan Brooke, at precisely 10:00, and went to bed. It had been Montgomery's habit throughout the war to retire promptly at 10:00 P.M., regardless of the battle situation, and the fact that he was playing host to the prime minister of Great Britain and the chief of the Imperial General Staff did not alter his life-style.

Churchill and Brooke were too excited to sleep on this eve of such momentous events. They went outside and walked in the moonlight, reminiscing of those dark earlier days when England, standing alone against Hitler's armies, had hovered on the brink of crushing, total defeat.

After Brooke retired to his trailer for the night, he scratched an entry in his diary as was his custom:

> It is hard to realize that within 15 miles hundreds of men are engaged in death struggles along the banks of the Rhine, whilst hundreds more are keying themselves up to stand up to one of the great trials of their lives. With this thought in mind it is not easy to lie down and sleep peacefully.[2]

Earlier that evening, British artillery opposite the town of Rees stepped up its rate of fire against German positions. At 9:00 P.M. the barrage reached an ear-splitting crescendo as the shadowy forms of men in the 51st Highland Division of General Brian Horrocks's XXX Corps began stealing toward the west bank of the Rhine. Looking grotesque in their bulky life jackets, the Highlanders scrambled into Buffaloes[3] at the water's edge and shoved off. In seven minutes the veteran British fighting men touched down on "Hitler's side" of the Rhine. Opposition had been surprisingly minimal. Men of General

Eugen Meindl's II Parachute Corps were scattered along the far shore, but had been pinned down and disorganized by the mammoth British artillery barrage.

The crossing at Rees by the Highlanders was a diversionary move designed to draw German attention away from Wesel. An hour after the Rees crossing, the 1st Commando Brigade, wearing green berets instead of steel helmets, climbed into Buffaloes two miles from Wesel. A brigadier of the Commandos, caught up in the emotion of the moment, exclaimed, "Hannibal crossed the Alps with elephants. We're making history crossing the Rhine with buffaloes!"

Halfway over the Rhine, the Buffaloes crammed with Commandos came under machine-gun and artillery fire. One of the craft was hit. Its gasoline tank spurted fire, sending crimson reflections across the water. Another Buffalo zigzagged to escape the enemy tracers, but it too burst into flame, and Commandos began leaping overboard, saved by their life jackets. The Commandos scrambled up onto the far shore and moved inland. Corporal Fred Taylor remained behind to stack ammunition when, by the light of a flare, he spotted the dim outlines of three German soldiers hurrying toward him. Taylor had put his rifle aside, so he began making threatening motions with his bayonet. The three Germans threw up their hands and began shouting, "*Kamerad! Kamerad!*" Only when the Feldgrau reached him did Taylor notice that it was not a bayonet he had been brandishing. In his excitement he had been waving a stick.

Brushing aside scattered pockets of resistance, the Commandos moved forward to the outskirts of battered Wesel, where they paused just as the drone of a large number of airplane motors reached their ears. At 10:30 P.M., 203 Royal Air Force heavy bombers began dropping 1,000 tons of high explosives on Wesel. Hardly had the last bomber turned and headed for England than the Commandos began edging into the pulverized city. The British fighting men appeared as eerie, ghostlike figures against the backdrop of fires raging out of control in Wesel.

Burrowed into the piles of rubble were members of a German force called the Wesel Division. The unit had hurriedly been thrown together only days before around a nucleus of antiaircraft artillery batteries whose members now found themselves engaged in desperate battle as infantrymen. Lashed on by their commander, Major General Friedrich Deutsch, the Germans fought tenaciously for every yard of rubble. Deutsch himself was in the forefront firing a Schmeisser machine pistol until he fell with a Commando bullet through his head. It would require 30 hours of bloody fighting before Wesel was secure in British hands.

South of Wesel at 12:30 A.M. on March 24, grim-faced soldiers of Easy Company of Leland Hobbs's 30th Division materialized out of the cellars and shadows and shuffled down the road single file toward the Rhine. They would cross in the first wave of assault boats.

A soldier near the front of the column called out, "All I can say is I hope those bastards on the other side know they're supposed to be *losing*."[4] As he spoke, 1,250 guns, plus those on tanks and tank-destroyers and hundreds of mortars, cut loose along Simpson's Ninth Army sector, engulfing the nighttime landscape in enormous echos of sound. No one in Easy Company spoke after that.

Over the shoulders of Captain Warme Parker's dogfaces shone the full spring moon. Ahead of them, on the far side of the Rhine, shell-bursts painted the skyline red and silhouetted them, distorted figures in their life-preservers. They trudged across open fields to the river, fields that stank of gunpowder and rotten cabbages. They walked through solid waves of thunder from their own artillery, which shook the marshy earth beneath their combat boots.

In fits and starts the men of Captain Parker's Easy Company reached the riverbank, where engineers in assault boats were waiting nervously. Crouching in the darkness, the Americans waited for H-hour and watched the far bank being pounded. Five more minutes. Three more. One. At 2:00 A.M. the assault boats, with 15 infantrymen crammed inside each one, shoved off.

At the tiller of one boat, engineer Private Chester Dabrowski of Minneapolis fiddled with the outboard motor. It sputtered and died. Muffled curses from 15 mouths. The men started paddling. The motor caught, sputtered again, then began running smoothly. As the boat headed across, it was blanketed by a smoke screen and the moon disappeared. On either side came the purr of 55-horsepower motors racing at 25 miles per hour through the smoke. In the front of Dabrowski's craft, Lieutenant Stanley Das was anxiously peering through the impenetrable fog for the first sight of the far shore. He expected to crash into other motor-powered assault boats at any second.

Suddenly the pall cleared. The moon reappeared. So did the other boats—and the enemy-held shore. The craft headed in, the motor was cut, and Das and the others could feel the scraping sensation as the boat edged onto the shore. A few Schmeissers pierced the night air as the Old Hickories scrambled out of the assault boat and trotted toward a dike 100 yards from the water's edge. "Hey, take it easy, guys," Corporal Willard Sorbert, loaded with heavy combat gear, yelled after them. No one turned to answer. They just kept going at full speed.

Captain Parker's men reached and scrambled over the dike past Germans dazed by the enormous barrage, who stumbled out of a stone cistern crying "*Nichts. Nichts. Kamerad.*" No one bothered to take them prisoners. The GIs kept racing ahead. "Hey, slow down, slow down!" several dogfaces called out. "It's a hell of a long way to Berlin."[5]

Earlier that evening, as one company of Old Hickories was trudging toward the Rhine, its tense members became aware that a figure had slipped out of the shadows and was marching in their ranks. The newcomer seemed out of place. He was wearing an officer's cap with a bill and the olive drab uniform, complete with necktie, normally associated with rear area personnel. Few were aware that Supreme Commander Dwight Eisenhower had fallen in step with them.

The general walked along with a young soldier who seemed especially fearful and depressed. "How are you feeling son?" Eisenhower asked.

"General," the youth replied, "I'm awfully nervous. I was wounded two months ago and just got back from the hospital yesterday. I don't feel so good."

As shells crashed along the far bank of the Rhine up head, Eisenhower replied, "Well, you and I are a good pair then, because I'm nervous too. But we've planned this attack for a long time and we've got all the planes, the guns, and the airborne troops we can use to smash the Germans. Maybe if we just walk along together to the river we'll be good for each other."

"Oh," the soldier responded, "I meant I *was* nervous. I'm not any more; I guess it's not so bad around here."[6]

Within two hours of crossing, the first line of settlements in the 30th Division zone had been cleared, and all three battalions had at least two battalions and tanks on the east bank. Total casualties for all three Old Hickory regiments had been fewer than those sustained by one regiment of Red Irwin's 5th Infantry Division of Patton's Third Army when it had sneaked across the Rhine at Oppenheim 48 hours earlier. Hundreds of surviving Feldgrau, covered with masonry dust, bleeding from ears, noses, and eyes, many speaking incoherently, had been incapacitated by the 65,261 Ninth Army shells that had exploded in their midst, projectiles ranging in weight from 25 to 350 pounds.

At the same time that the American Old Hickories were crossing Germany's historic barrier to invasion from the west, elements of Lieutenant General Neil Ritchie's British XII Corps, spearheaded by the Scottish 15th Division, were being ferried over the Rhine northwest of Xanten and within minutes touched down on the far shore. Only sporadic artillery, mortar, and machine-gun fire opposed the Scots' crossing.

At 3:00 A.M., one hour after Hobbs's 30th and Ritchie's Scottish 15th Divisions had stormed across, General Ira Wyche's U.S. 79th Infantry Division headed for the east bank south of Hobbs's crossing. The wind had diminished, and a mixture of nature's fog and man's smoke clung to the river and to both banks, causing the formation of assault boats to scatter during the 500-yard run. Confusion reigned. Men in some craft lost direction and went ashore on the west bank, from which they had departed 20 minutes before. Thinking that they had landed on the German-held side of the Rhine, men in one boat quickly formed into a skirmish line and charged inland, only to come face to face with an American company headed for the riverbank. A disastrous firefight was narrowly averted.

On the east shore, the units became intermingled, but in the ab-

223

sence of heavy German resistance, the assault battalion organized rapidly and pushed inland, past German soldiers too stunned from the artillery bombardment to engage them.

Dawn broke bright and clear. There would be no respite for the reeling Wehrmacht along the Rhine, as the greatest coordinated aerial strike in history was unleashed by the Anglo-Americans. Between daylight and dusk, 10,109 American and British warplanes would pound the German army and its facilities. High in the sky, their fuselages glinting in the early morning sun, thousands of U.S. Eighth Air Force and Royal Air Force heavies left white vapor trails as they headed for Wehrmacht rail and communications centers at Damstadt, Giessen, Fulda, Frankfurt, and other cities.

At the controls of a Mustang fighter, Major General Hoyt Vandenberg, the youthful leader of the U.S. Ninth Air Force, flew over the region east of the Rhine for three hours to direct his 1,806 fighter-bombers and 812 mediums as they made life miserable for the Germans. Forty-year-old Elwood R. "Pete" Quesada, boss of the Ninth Tactical Air Command, flew his squat Thunderbolt up and down Montgomery's sector of the Rhine to oversee attacks by his 1,007 fighter-bombers as they strafed and bombed. Quesada radioed back: "I see a hell of a lot of Krauts streaming back toward the Rhine with their hands in the air."

That morning at Third Army's bridgehead at Oppenheim, marching men, field artillery, tanks, and trucks had been pouring over pontoon bridges. An epidemic of bladder trouble seemed to have broken out among Patton's soldiers; a large number of them were urinating off the bridge just as they reached the center of the Rhine.

"The first thing I did was to let a big one go right in the river," Private Morris Berkowitz, a rifleman from the Bronx, told a comrade.[7]

"You take the Rhine," suggested Anthony Marino of Boston. "I'll take the Charles River back home."

Talk was suddenly cut off as the thump-thump of several score of Third Army antiaircraft guns erupted. American machine guns sent up streams of tracers, and the men crossing the bridge looked skyward to see five Messerschmitt fighter-bombers flying toward the spans. One of the Luftwaffe planes exploded in an orange burst of fire as a 90-millimeter shell caught it broadside. The other pilots pulled up, banked, and headed for their home base.

"We're leading the league again!" crowed Private Chick Johnson of New York City, a member of the all-black 452nd Antiaircraft Battalion. "We just hit another home run!"

"Yeah," responded a comrade. "And we'll get a chance to bat a lot more. Seems like the Luftwaffe is coming in every minute."[8]

A short time later a jeep adorned with a red metal pendant bearing the three gleaming stars of a lieutenant general edged onto the pontoon bridge at Oppenheim, headed for the east bank. General George Patton, ramrod straight in the front seat, was crossing the Rhine for the first time. Hunched together in the small back seat were his two longtime aides, Lieutenant Colonel Charles Codman, a peacetime wine merchant, and Major Alexander Stiller, a Texan whose job primarily was that of bodyguard to the Third Army commander.

Nearing the middle of the river, Patton called out, "Hold up! Time for a short halt." Without a word he got out of the jeep and walked to the edge of the bridge, peering down briefly at the swift-flowing current. He unbuttoned his trousers and performed the same rite that Winston Churchill had carried out on the Siegfried Line a few days previously.

Returning to the jeep, Patton exclaimed joyfully, "I've been waiting a hell of a long goddamned time to do that."[9]

When the staff gathered at General Omar Bradley's headquarters that morning of March 24, Bradley was in a good humor. So were his aides. Only a short time before they had heard a prerecorded Churchill statement eulogizing the British in general and Bradley's rival Bernard Montgomery in particular for conducting the "first assault crossing of the Rhine in modern history."

"First, hell!" someone called out. "Hodges and Patton are already over. Monty's come in third." A roar of laughter greeted the remark.

General Bradley was still chafing from Montgomery's January 7 press conference after the Bulge, and from the fact that the U.S. Ninth Army was riding again under Monty's banner—albeit anonymously. So Bradley could hardly restrain his glee when Major Stiller, Patton's aide, presented a brief oral report on Patton's sneak crossing at Oppenheim, without mentioning Montgomery by name:

"Without benefit of airborne drop, United States or British navy, and not having laid down the greatest smoke screen in history, and without a three-month buildup of supplies or a whole extra American army, and with no preliminary bombardment, and finally without even a code name, Lieutenant General George Smith Patton and Third Army at twenty-two hundred hours [10:00 P.M.] Thursday evening, March 22, crossed the Rhine."[10]

In their euphoria over beating Field Marshal Montgomery over the Rhine, Bradley and his commanders had overlooked one significant factor: Ultra had told George Patton that the Germans had virtually no forces on the far bank facing him, whereas Field Marshal Kesselring had congregated some of the best troops still available to the Wehrmacht—mainly parachute and panzer units—to try to fend off Montgomery's great blow.

20

The Paratroopers Strike

Before dawn on March 24—D-day for Operation Varsity—11 airfields in southern England were astir with activity. Red Devil paratroopers and glidermen were feverishly at work. Most found it hard to believe that in less than five hours they would be whisked out of this pastoral English countryside and dropped into a German inferno appropriate to their division's name.

As would the American 17th Airborne Division, General Bols's 6th Airborne would have a full complement of correspondents along to cover the historic first attack from the sky east of the Rhine. Doon Campbell of Reuter's was delighted to find that he had been assigned to a glider which had a jeep aboard. The jeep would come in handy when he landed. His elation turned to concern when he found that the jeep was loaded from top to bottom with hand grenades.

"Say," Campbell asked a Red Devil vet of previous combat assaults, "what happens to a glider loaded with hand grenades when it's hit with enemy fire?"

"No need to worry, old man," was the cheerful reply. "You'll never know it."

Robert Vermillion of United Press, an American, and Geoffrey Bocca of the London *Daily Express*, were also going along. Bocca was very young, very soft-spoken, eager to help his fellow man, and slightly apprehensive about his first combat mission. Vermillion had jumped into southern France in August 1944 and had allowed his typewriter to be shoved out in an equipment bundle. He had had a difficult time running the machine down. As Vermillion would parachute again and Bocca would be floating down in a glider, the American prevailed upon his young British colleague to take the typewriter with him.

Across the English Channel from Red Devil encampments, members of Bud Miley's 17th Airborne Division began scrambling out of tents before dawn. Most had been sleeping fitfully. The big moment in their lives was only hours away. Breakfast was served. Most men only picked at their food. There was little talk. The Thunder from Heaven troopers filed past to pick up their parachutes. Now all there was to do was wait for the dreaded order all knew had to come— "Load 'em up!"

Miley's men were given the option of wearing a reserve chute. Private First Class John Magill, the Pennsylvania farm boy, strapped one on. Flying in at low altitude, Magill knew there would not be time to use the spare chute if his main chute failed to open, but he felt it would be psychologically comforting to feel the reserve on his chest. Besides, it could be a shield against flak or small-arms fire on the descent to earth. Many others felt the same way.

At an airfield in the Paris region with the unromantic designation A-58, glider and C-47 pilots of the U.S. 437th Troop Carrier Group were jeeping the one mile from the garrison area to their aircraft. Most were solemn-faced and sleepy. Few had slept during the night. Instead, they had kept busy by telling tall tales and making final preparations. Who could, or would want to, sleep? This might be their last night on earth.

The C-47s and the gliders they would pull had previously been placed in position, so with a beautiful spring day breaking there was little for the nervous pilots to do but wait. They glanced periodically at their watches. Moments later they would look at them again. They tried not to think of the horrors they had witnessed on previous flights—the flaming crashes of the flimsy gliders and the lumbering C-47s in Normandy, in Holland, and in resupply missions over Bastogne in the Battle of the Bulge. They tried to block out the specter of these crashes, which had taken buddies to their deaths, but it was hard not to remember.

In recent days ground personnel of the 437th Group had rushed to an army ordnance depot near Paris where a large number of wrecked armored vehicles were stored. They cut off big chunks of armor plate, which were placed in the pilot and co-pilot seats of the C-47s and gliders. This at least provided a psychological boost.

"When the flak and the machine-gun bullets are coming up at you, your ass feels awfully naked," a 437th pilot summed up. That was the view of his comrades who would fly elements of the 194th Glider Infantry Regiment over the Rhine.

The flight plan for Operation Varsity called for the 6th and 17th Airborne Divisions to rendezvous over Brussels, Belgium, after Bols's Red Devils took off from England and Miley's Thunder from Heaven men lifted off from the Paris region. Once the long streams of C-47s and gliders joined at Brussels, the entire flight would turn northeast for the final 100-mile run to the DZs (drop zones) and LZs (landing zones) four to six miles east of the Rhine River. As the 200-mile-long air armada, which would take two hours and 37 minutes to pass a given point, neared the river, all American and British artillery fire was to cease. Allied airborne planners were taking no chances of a repetition of the disaster that had occurred in the invasion of Sicily in July 1943. At that time, rattled and confused American navy and army gunners had opened fire on a flight of low-flying C-47s carrying members of the 82nd Airborne Division, shooting down 23 of the transport planes and killing more than 300 American paratroopers.

As the time for lifting off neared, General Bud Miley was going from plane to plane at a dusty airfield near Châlons-sur-Marne, giving encouragement and exhorting his men to fight with vigor and determination. The soft-spoken, 48-year-old Miley was not the fire-eating brand of airborne leader, but rather led by example. He enjoyed great respect in all ranks of the 17th Airborne. Calmly going about his task, Miley gave no indication of the emotion he felt. For years he had been preparing himself for this moment—leading an airborne division in a combat jump behind enemy lines.

Bud Miley had been born into the army; his father had been an 1887 graduate of the U.S. Military Academy at West Point. A brother, John, Jr., had received a commission at West Point in 1916, and Bud graduated from the academy two years later. In 1940, Major Bud Miley took command of the 501st Parachute Battalion, a fledgling organization, and became the first U.S. Army officer to lead a designated unit of airborne troops.

A star athlete, Miley took quickly to airborne training and was credited with establishing the unwritten rule among parachute and glider troops that an officer would not ask a subordinate to undergo any risk which he himself would not take.

As a major general, Miley took command of the 17th Airborne when it was activated on April 17, 1943, at Camp Mackall, North Carolina, a desolate chunk of countryside rife with sand dunes and pine trees. He led it overseas late in 1944, and the men of the division settled down for a peaceful and cozy Christmas in England. Suddenly, Miley's eager but green airborne men were rousted from their beds, rushed to the Continent, and thrown into the Battle of the Bulge as straight infantry. There they had helped halt the German breakout.

Dawn was breaking when the first aircraft carrying the 4,876 men of Bols's 6th Airborne Division streaked down runways in East Anglia and lifted into the sky. Just before the departure, one of Hitler's "vengeance weapons," a buzz bomb apparently targeted for London but far off course, crashed into a field near one of the aerodromes. The enormous concussion shook the transport planes on the runways but did not affect the departure schedule.

"Ol' Adolf must have hoisted a few too many last night, and his bloody aim was poor this morning," a Red Devil observed to comrades.

The Red Devils would have longer to journey than would the 17th Airborne in France, so they took off first. A long swath of southern England shuddered beneath the constant throbbing of engines as 429 British planes and 247 C-47s of the U.S. IX Troop Carrier Command headed for battle with cargoes of paratroopers.

At about the same time that Allied aircraft were lifting off from England, men of the 507th Parachute Infantry Regiment and attached units were climbing aboard C-47s at several airfields in the vicinity of Chartres, France. Some members of the 507th had jumped into Normandy when the regiment had been a component of the 82nd Airborne Division. But the 507th had suffered heavy casualties in savage fighting there, and its ranks were filled with men going into a combat jump for the first time.

Flying in the lead plane would be the regimental commander, Colonel Ed Raff of New York City, who would be the first American to bail out east of the Rhine. On the field, as his grim-faced men strapped on their parachutes and loaded themselves down with more than 80 pounds of combat gear, the scrappy Raff raced from aircraft to aircraft. At each plane the colonel's jeep would stop. Raff would stand up in the vehicle, shake his fist at the knot of paratroopers waiting to climb aboard, and shout, "Give the goddamned bastards hell, men! You know what to do. Cut out their goddamned guts!"

Now there were shouts of "Load 'em up!" and men of the Thunder from Heaven division, burdened by heavy gear, waddled into line and began filing up the short ladders and through the doorways of the C-47s. Colonel Raff turned to Private First Class Cleo I. Crouch, a 29-year-old trooper from St. Joseph, Missouri, whom the colonel had named as his bodyguard a few days before. A commander, often occupied with studying maps and talking on field telephones in tense combat situations, needed a tough, resourceful fighting man for protection against a surprise assault. Crouch and his Tommy gun would provide that protection.

"Crouch," Colonel Raff reminded him curtly just before climbing into his C-47, "I'm going to be damned busy when we get over the river. If anybody even *looks* like they're trying to harm me, shoot 'em!"

"Yes, sir," the tall, thin bodyguard replied. Soon to make his first combat jump, Crouch pondered fleetingly over who would protect him while he was protecting the colonel.

At 7:17 A.M. Colonel Raff's transport plane, carrying his headquarters staff, lifted off and set a course for Brussels. Other bulky C-47s took off at brief intervals. Soon to follow would be the 17th Airborne's other parachute regiment, the 513th and then the 194th Glider Infantry Regiment. General Miley would be among the first paratroopers to jump.[1]

It was 8:58 A.M. when the last plane took off. The noise of the armada reverberated across the French terrain for miles. Civilians gazed up in awe as the seemingly endless stream flew past. The enormous column extended farther than the eye could see—610 C-47s towing 906 gliders, 72 C-46s and 226 C-47s, carrying 9,387 men of the 17th Airborne Division.

Flying northeast through the bright spring sky, the enormous sky train rendezvoused with the British stream southeast of Brussels, and the two columns, flying side by side, set a course for the drop and landing zones at Wesel. Darting about like mother hens protecting their broods were 676 fighters of the U.S. Ninth Air Force and 213 fighter planes of the Royal Air Force.

Seated next to Colonel Raff was Sergeant Harold E. Barkley, a bazooka squad leader. Barkley was one of the regiment's old men—all of 30. Trying to appear as calm as possible although his stomach was churning, Barkley glanced around at his comrades seated tensely in bucket seats in two rows facing each other. The faces he saw were young, pale, and drawn. Here and there lips moved almost imperceptibly in prayer. This was the first combat jump for most, including himself. Barkley presumed that his own face looked pale and tense also.

Raff, a hard-bitten fighting man, started to fidget in his seat when the sky armada had been in flight for nearly two and a half hours.

Operation "VARSITY"
TROOP CARRIER ROUTES

Wesel

Essen

Dusseldorf

17 Div LZ DZ   6 Br Div LZ DZ

Its South tip

Its North tip

Antwerp

Waavre

Brussels

Dunkirk

Calais

Airfields for 3 & 4 46 Groups
RAF and 3 groups of 50 Wing

52 WING
AREA

Abbeville
Amiens
Poix

Achet

Clery

52 WING
AREA

Provins

Mormeaux le Grand

Coulommiers

Chateau Thierry

53 WING
AREA

Meaux

PARIS

Bretigny

Chartres

Etampes

CHATEAUDUN
AREA

Dreux

Chateaudun

50 WING
& 17 P.G.
AREA

Evreux

LONDON

GREAT DUNMOW

WETHERSFIELD
BOREHAM
CHIPPING ONGAR
MATCHING

ENGLISH CHANNEL

0   50   100

Operation Varsity Troop Carrier Routes

233

Staggering a little under his heavy gear, Raff went forward to the cockpit to talk with the pilot, Colonel Joel L. Crouch, leader of the Troop Carrier Command pathfinder group whose job it was to guide the formation to the drop zone. "I'll bet you a case of champagne we drop you right on the button," Crouch declared.

"You're on!" Raff fired back.

Raff had led the 509th Parachute Infantry Battalion when it made the first American combat jump outside Oran, Algeria, in November 1942, during the invasion of North Africa. Crouch had been a pioneer in developing pathfinder techniques for troop carrier aircraft and had flown the lead plane in combat missions to Italy, southern France, Normandy, and Holland. Now it seemed fitting that these two pioneers, Colonels Raff and Crouch, held places of honor in the van of what promised to be the final major airborne operation of the war.

Unseen some distance ahead of the air convoy, Winston Churchill, General Eisenhower, and Field Marshal Brooke were standing on a hill overlooking the Rhine, a mile south of Xanten. It was 9:30 A.M. The Allied leaders had come to view one of the largest military spectacles in history. They were destined for disappointment: it was so hazy and smoky on the far side of the Rhine that visibility was limited to less than a mile.

At 9:51 A.M. Colonel Raff's plane was nearing the Rhine. Behind him were the 181 C-47s carrying the 2,479 men of his 507th Parachute Infantry Regiment and its teammate, the 464th Parachute Field Artillery Battalion, commanded by Lieutenant Colonel Edward S. Branigan, Jr., of Manhasset, New York. Suddenly the red light flashed on in the cabin of Raff's plane—four minutes to bailout. The red glow sent a new chill through the paratroopers. Above the roar of the engines and the wind rushing through the open jump door of the C-47, Colonel Raff called out, "Stand up and hook up!"

The thump of antiaircraft guns could be heard to the front. There was a rustling of gear as troopers struggled to their feet. Knees felt like jelly, partly from the heavy burden of equipment, partly from extreme tension. Mouths went dry as cotton. Some men felt sick.

Ed Raff crouched in the door, his hands gripping each side. He felt a surge of excitement on seeing the Rhine, a shimmering silver strip.

234

Beyond the wide moat was a thick pall which veiled the drop and landing zones. Added to the natural early morning haze were the lingering smoke from Field Marshal Montgomery's generators, which had shielded the amphibious assault troops during the night, and clouds of dust blowing over the region from heavily bombed Wesel.

Troopers in Raff's plane stared at the red light next to the open door as though mesmerized. Why did it have to be red? Red spelled danger. Suddenly, the light turned green and Colonel Raff shouted, "Let's go!" and jumped. Right on the regimental leader's back, almost literally, was Sergeant Barkley, the bazooka squad leader, followed at split-second intervals by others in the stick. The cabin was cleared in less than 10 seconds.

Paratroopers—hundreds of them—were spilling out of other planes, and in seconds the hazy sky east of the Rhine was filled with billowing white parachutes. Barkley felt his parachute pop open. He was surprised that no one on the ground was shooting at him. As he floated downward, Barkley became aware of a large barn directly below him. The structure's shingles had been blown off, and all that remained of the roof were rafters and sheeting. There wasn't time for the sergeant to slip to one side, so he crossed his feet, grabbed his risers, and uttered a quick prayer. Barkley fell through a two-foot gap between the rafters and landed unhurt in the barn's mow filled with loose hay. Down below in the barn he heard a rustle. Gripping his rifle, the sergeant peered over the edge of the mow. Staring curiously at the intruder were a pair of wide-eyed cows.

Private First Class Gorden L. Nagel jumped from a C-47 carrying a 22-pound machine-gun barrel. As his chute popped open, giving his body a terrific jerk, the heavy barrel struck him across the face, breaking his nose and causing blood to pour into his eyes. He was blinded. He could hear ack-ack guns around him and the ripping noise of bullets piercing his parachute—sounds he had come to know the previous June when he had jumped into Normandy.

Nagel was fearful that he would break his legs or back on striking the ground, because he could not see it to brace himself for the shock. To his surprise he touched down in a routine manner, then rolled over and shucked his chute using the quick-release harness being tried in combat for the first time by the 17th Airborne. Nagel could hear firing around him as he frantically labored to stem the flow of blood from his nose and get his vision back.

Slowly his sight returned and he headed for the sound of heavy firing. A short distance away he came upon two close friends, one an outstanding boxer who Nagel had been convinced would one day be world heavyweight champion, and the other a youth who was a skilled artist. Their chutes had caught on high trees and they dangled from harnesses. They had been riddled by bullets. Both were dead. "And they never even had a chance to fire a shot!" Nagel anguished.

Lugging his machine-gun barrel, Nagel trudged on. Overhead, other waves of C-47s were discharging their paratroopers, and he could see thick puffs of smoke where shells exploded around the lumbering aircraft. One C-47 disintegrated in a fiery ball. Minutes later the machine-gunner came upon a medic feverishly working over a parachutist stretched out on the ground. Nagel recognized the wounded young trooper—another close friend, who was always laughing and joking. Now he lay still, his face ashen.

"Can't understand it," the medic said. "It looks as though he got a bullet right through the heart, but he's still alive."

The youth died seconds later. In only minutes after landing, Nagel had lost three close buddies. "What are my odds for surviving this mission?" he asked himself. He hurried off in search of his company.

Approaching the Rhine River in a C-47, Private First Class Robert M. Baldwin, a rifleman from Ridgway, Pennsylvania, had more than a paratrooper's customary worries as he stood hooked up and ready to jump. Only a few hours before the 507th Parachute Infantry had taken off, Baldwin had rejoined his company from the hospital after an extended bout with diarrhea which, he told comrades, he had "lost by a knockout." At a time when he needed all the physical stamina he could muster, Baldwin felt weak and shaky. He pondered whether he would be able to handle the exertion of the jump and the subsequent action.

Equally worrisome was the fact that Baldwin had rejoined his unit on such short notice that he had not had time to obtain his own rifle, and instead was going into battle with a "strange" Garand. He did not know if it would even shoot. And he had no chin strap to brace his head when his parachute popped open violently; he conjured up the specter of his neck snapping from the heavy jolt.

Suddenly, the red warning light in Baldwin's C-47 turned to green, and men started bailing out. Baldwin was the fourth man through the opening. Seconds after his chute blossomed, a bright orange flash erupted just below him. His platoon sergeant had been in that parachute, and now he was not there. A German bullet had hit a bag of plastic explosives the sergeant had been carrying, and in the subsequent explosion he simply vanished into powder. The victim's chute continued downward, reminding the horrified Baldwin of a racehorse galloping away after having thrown its rider.

Bullets were hissing past Baldwin as he descended. They seemed to be coming from all points on the compass; he was convinced that the entire German army was out to get him personally. Landing in a large field between a shed and a large bomb crater, Baldwin got out of his chute and leaped into the good-sized hole head first. There was a squishing noise as he landed directly upon a gruesomely mangled German corpse. Bits of what had once been a human being were all over Baldwin.

Despite the enemy machine-gun bullets, Baldwin couldn't stay where he was. Flicking pieces of bloody intestines from his face and jumpsuit, the paratrooper scrambled to his feet and dashed toward a seven-foot-high fence some 40 yards away. German gunners spotted him and fired as he reached the tall obstacle. Somehow he pulled himself over it.

Diving into a nearby ditch and panting from the exertion, Baldwin glanced back at the fence. How had he gotten over it in his weakened condition? The answer promptly struck him: he had been "inspired" by the German bullets.

21

Dropping onto a Hornets' Nest

Circling the drop zones in the Wesel area was a four-motored Flying Fortress carrying American correspondents and other observers. Among those taking in the panorama was Richard C. Hottelet of CBS Radio. Hottelet had never seen such a vast operation. Neither had another observer in the B-17, Colonel Joel O'Neal of San Antonio, Texas.

"That's the most concentrated jump I've ever seen," O'Neal exclaimed. "And I've seen a lot of them."

Below the Flying Fortress was a kaleidoscope of white, red, blue, and yellow parachutes, the different colors representing different items of equipment, ammunition, medical supplies, or weapons. After the final paratrooper had jumped, the pilot called out, "I'm heading back to pick up the gliders. They should be approaching now." He banked the big aircraft and started over the Rhine. Hottelet and the others quickly picked up the column of gliders and tugs heading toward them, the tail of the stream disappearing into the blue many miles in the distance.

Suddenly 20-millimeter shells from German guns on the ground began plowing through the B-17. The left wing and a motor caught fire, and smoke poured out. Hottelet felt his heart skipping a beat. The pilot calmly advanced his throttles and pulled the ship up a couple of hundred feet. The motor burned fiercely. Then another engine caught fire. Hottelet did not relish the thought of bailing out, but his alternative was to ride the doomed ship down. He rapidly checked his parachute, not certain what it was he was checking for.

In an even voice, the pilot called out, "We're at six hundred feet. Get ready to jump!"

Hottelet had never jumped before, but with the cabin filled with smoke and two engines in flames he needed no more urging. Colonel O'Neal went out first, with the CBS Radio war correspondent right on his tail. As Hottelet's chute opened, he received the customary terrific jolt. His chest pack bounced upward and struck him across the face, causing him to temporarily lose his vision in one eye.

O'Neal and Hottelet landed without difficulty in a large green field near a British artillery battery some 2,000 yards west of the Rhine. A British captain drove up in a few minutes and offered the two Americans whiskey, which they eagerly accepted. Off in the distance, they saw six C-47s crash in flames.

Flying over the haze-covered DZs in another aircraft, this one a C-47, was Major General James Gavin, leader of the 82nd Airborne Division. His outfit was being held on the alert back in France, not to reinforce the Rhine operation but to parachute into Berlin—Operation Eclipse. Gavin had led his paratroopers in four combat jumps, and with no immediate assignment and his curiosity piqued, he decided to fly up to watch Operation Varsity as a spectator.

Knowing that he held the top-secret knowledge of the looming Berlin airborne strike and that he would be a highly valuable prisoner should he be shot down in the Wesel region and captured by the Germans, Gavin instructed his pilot to circle at about 1,800 feet and not to stray too far from the Rhine. This altitude would be far above the 600 to 800 feet at which the troop carriers would fly.

"At eighteen hundred feet we won't be of much interest to the Germans," Gavin had reassured his pilot. "They'll have plenty of targets much lower."

After circling for a half hour, Gavin's plane suddenly drew the attention of German antiaircraft batteries. Sharp cracks were heard as shells exploded nearby. The C-47 rocked from the concussions. There was a curious sound, which the airborne general had heard in previous operations, as though a small boy had tossed a handful of pebbles against the tight skin of the aircraft—shrapnel.

"They're getting our range," Gavin calmly told the pilot. "We better get on home. I'm not even supposed to be here."[1]

As the C-47 banked and winged back across the Rhine, General Gavin turned for one final look. He saw several troop carriers going down in flames. God, he thought, I hope the troopers and crews got out in time.

General Miley, in the meantime, had bailed out with key staff members, and on landing was immediately raked with machine-gun fire. Pinned to the ground with bullets passing just overhead, the division commander glanced around for those who had jumped with him. He could not spot a single staff officer. A short distance away, Miley saw three privates clinging to the ground to escape the machine-gun bursts, and to one side the general saw an equipment bundle with markings that indicated it contained a machine gun.

Above the chatter of German machine guns, General Miley called out to the three troopers, "Meet me over at that bundle." He pointed in the direction of the disassembled automatic weapon and its ammunition. Miley and the three men slithered over the ground to the bundle, ripped it open, and assembled the gun. Within minutes the general and the trio of privates were engaging the German gun.[2]

When firing diminished in the immediate area, Miley departed in search of his command group. When he was a short distance away one private at the machine gun nodded in the direction the general had taken. "We sure as hell had some goddamned high-priced hired help today."

Throughout the areas of the drop zones, countless vicious small firefights had erupted. Private George J. Peters, a soft-spoken, black-haired platoon radio operator, a native of Cranston, Rhode Island, had parachuted into a field near the town of Fluren, along with 10 other troopers. About 75 yards away, a German machine gun, supported by a squad of riflemen, opened fire on the parachutists, pinning them to the ground. Caught in the open as they struggled to free themselves of parachutes, Peters and the others appeared doomed. Suddenly, and without orders, the mild-mannered Peters leaped to his feet and, armed with only a rifle and a few grenades, began a one-man charge against the German strong point. Spotting the lone baggy-pantsed American racing toward them, the enemy gunners concentrated their fire against him.

Firing as he ran, Peters was halfway to the spitting machine gun when a burst from the weapon knocked him down. Bleeding profusely, he struggled to his feet and continued the charge. Again bullets plowed into his body and he went down for the second time, unable to rise. Crawling forward until he was only some 15 yards from the German position, Peters pitched two grenades, then collapsed. A fiery blast erupted at the machine gun, killing the enemy crew and causing the riflemen to flee. Moments later George Peters was dead.

Colonel Raff, meanwhile, had landed under fire, and after shucking his chute took bearings of the situation. He and all of his troopers were to have been dropped in a large clearing two miles north of Wesel at the lower end of the Diersfordter Forest; but the first serial of troop carriers had dropped the regimental commander and 493 of his parachutists two miles northwest of their DZ, near the small town of Diersfordt. The remainder of Raff's regiment, including Lieutenant Colonel Branigan's 464th Field Artillery Battalion, landed directly on or adjacent to its designated drop zone.

Proudly calling themselves Branigan's Bastards (due to the fact that they were not an organic part of the 17th Airborne Division), the artillerymen were hit by heavy machine-gun fire from a nearby woods as they parachuted down. Bullets came so close to Branigan as he drifted down that his dispatch case was riddled. Slithering over the landscape under a hail of fire, the artillerymen hastily set up three .50-caliber machine guns and began firing back at the German automatic-weapons crews.

Meanwhile, other artillerymen were crawling around the DZ trying to locate and assemble the parts of the battalion's howitzers. They were stymied by heavy machine-gun and rifle fire from a farmhouse on the edge of the DZ. The howitzers' parts had been dropped in separate bundles and were scattered all over the drop zone.

Spotting the Germans in the farmhouse, Lieutenant Colonel Branigan rounded up five of his nearby men, including First Sergeant Edmund L. Kissinger. Pointing at the enemy strong point, Branigan said, "We're going to charge that farmhouse and wipe out the Krauts inside."

Firing their Tommy guns and rifles and tossing hand grenades, Branigan and his five men ran over open ground toward the farmhouse. Sergeant Kissinger went past the battalion commander, pitched a grenade, and fell dead, a bullet through his head. Branigan and the remaining paratroopers bolted inside the house, spraying the rooms with Tommy-gun fire. Those Germans not killed quickly surrendered.

Three howitzers were by now assembled, and their crews opened direct fire on the most persistent German positions. Branigan's Bastards had achieved a coveted goal: the first airborne artillery outfit to land, fight, and fire its guns east of the Rhine. After enemy resistance around them died down, the men of the 464th moved according to plan to the northeast end of the DZ and soon had 9 of their 12 howitzers in action.

In the meantime, Colonel Ed Raff had rapidly assembled the nearly 500 men who had landed with him off-target and, clutching his Tommy gun, set out for his regiment's drop zone two miles to the southeast. In his path was Schloss Diersfordt, a thick-walled fortress in the center of the Diersfordter Forest and a major objective of the 507th Parachute Infantry. Unknown to Raff, the medieval fortress was being used as a headquarters by a German corps. The colonel, itching for action, soon found it. At the edge of a woods a German force opened fire on the advancing paratroopers, but Raff's men drove the enemy from their positions after a brisk but brief firefight, killing 55 Germans and capturing over 300, including a lieutenant colonel.

Off in the distance, perhaps three-quarters of a mile away, Raff spotted a German battery of 155-millimeter howitzers firing against ground troops pouring over the Rhine six miles away. Instead of skirting the enemy battery, Colonel Raff told his men, "We're going to take the bastards." The paratroopers overran the guns and killed, captured, or scattered the crews. The howitzer barrels were spiked with thermite grenades, and the airborne men moved on toward Schloss Diersfordt.

Bob Krell, a reporter for *Yank*, a weekly GI magazine, was accompanying a leading patrol when suddenly, from ambush, German machine-gunners opened fire. Krell and the rest of the patrol were killed.

On the way to the castle, grim-faced men passed a comrade hanging from a tree in his parachute harness, a bullet hole in the center of his forehead. A painted white cross was on his helmet. The troopers recognized the dead comrade immediately—the highly respected Episcopal chaplain of the 507th Parachute Infantry.

Nearing Schloss Diersfordt, Colonel Raff heard a heavy firefight in progress. His 3rd Battalion, which had landed on the DZ, was already at the rambling old castle and putting its German defenders under rifle and machine-gun fire. Troopers were preparing to charge Schloss Diersfordt when two panzers clanked out of the castle and rolled down a narrow forest road toward the airborne men. A trooper fired an anti-tank grenade, which struck the first panzer, disabling it and causing the crew to surrender. A team of tank hunters, armed with the new 57-millimeter recoilless rifle, stalked the second German tank and set it afire with a direct hit. This time no one emerged from the burning panzer.

While two companies of the 3rd Battalion sent a fusillade into turrets and upper windows, Company G troopers bolted into the cavernous structure and for the next two hours fought the Germans from room to room. When the shooting ceased, the 300 surviving Germans were taken prisoner, including a number of colonels and majors on the staff of Major General Hans Straube's LXXXVI Corps.

In the dark basement of the castle, Raff's paratroopers rescued nine comrades who had been captured and locked up. While the battle raged outside and within the castle, the Americans had discovered a cache of wine and liquor stamped RESERVED FOR THE WEHRMACHT. When rescued, all nine captives were drunk.

As the American troopers, who sometimes called themselves Raff's Ruffians, were bailing out that morning, British General Eric Bols's Red Devils of the 6th Airborne Division were floating to earth directly onto their drop zones in the northern portion of the airhead. Bols himself was among the first to jump. Flak had been heavy since the sky train passed over the Rhine, and several troop carrier planes burst into flame and crashed.

On the way to the ground Bols's parachutists caught heavy machine-gun fire. Lieutenant Colonel J. S. Nicklin, commander of the 1st Canadian Parachute Brigade, dangled helplessly when his chute caught in a tall tree. A burst of automatic weapons fire riddled Nicklin, killing him almost instantly.

Furious over the death of their commander, the Canadians, despite heavy enemy fire, cleared their objective in less than two hours. So many Germans were taken prisoner that they became a problem—the POWs outnumbered the Canadians. The defenders were by now suffering heavy casualties. A Canadian staff officer radioed to higher headquarters: "Germans are being killed by the hundreds."

Corporal Frederick G. Topham, a medic with the Canadian paratroopers, was giving treatment to a group of wounded men when he heard a cry for help. It came from a comrade in an exposed position in an open field. Two medical orderlies dashed out to help the wounded man, but each was shot and killed as he knelt beside the casualty. Corporal Topham saw the two fellow medics gunned down. Without orders, he raced to the middle of the field as German machine-gunners sent streams of bullets hissing past him. As Topham hovered over the seriously wounded man, a bullet tore through the medic's nose, shattering the bone and causing blood to gush out. Despite the excruciating pain, Topham laboriously carried the injured Canadian through intense fire to the relative safety of a woods.

The young corporal returned to the open field time and again to carry wounded Canadians to safety, all the while being subjected to machine-gun fire. Only when he had all his injured comrades under cover did he consent to have his gaping nose wound treated. At the aid station, an officer ordered Topham evacuated. The medic protested so vehemently that the order was rescinded.

On the way back to rejoin his company, Corporal Topham came across a burning ammunition carrier loaded with mortar shells. It had received a direct hit by an enemy projectile, which had wounded three men still in the smoking vehicle. German mortar shells were exploding around the carrier. "Don't go near it," an experienced Canadian officer called out. "It'll blow at any second!"

Ignoring the order, Topham dashed out to the vehicle as mortar shells exploded around him, and dragged the three men out of the smoking vehicle. One died minutes later, but the other two were still alive. Topham hurried off to rejoin his company.

Brigadier James Poett's 5th Parachute Brigade was dropped near the town of Hamminkeln, where the Red Devils ran into tenacious resistance from dug-in Germans on the outskirts. After dropping the 1,917 men of Poett's brigade almost directly onto its DZ, the 121 aircraft of the U.S. 315th and 316th Troop Carrier Groups turned left and set a course for home. Almost at once, a torrent of antiaircraft fire was unleashed from the ground, and 10 airplanes were shot down in flames, some taking crewmen to their deaths. Seven others crash-landed in Allied territory, and an additional 70 were damaged, most of them heavily. In this stretch of only a few miles, 6 airmen were dead, 20 missing, and 15 wounded.

On reaching the ground, Brigadier Poett's men and those of Brigadier Hill's 3rd Parachute Brigade clashed at once with the German defenders. Throughout the heavy action, the sounds of hunting horns, a traditional British method for rallying troops, echoed above the din of battle. The haunting notes from the horns lent an eerie touch to the holocaust that had erupted on all sides.

Meanwhile, General Horrocks's British XXX Corps had run into a hornets' nest at the town of Rees after having crossed the Rhine during the night with surprisingly moderate opposition. By morning, Horrocks began receiving reports of Germans surrendering in large numbers to the American and British forces on each flank, but the Scottish 51st Highland Division signaled the corps commander that it was meeting the most tenacious opposition since the bloody fighting in the earthen hedgerows of Normandy.

German *Fallschirmjäger* (paratroops) and panzer forces were clinging to the rubble of Rees and appeared intent on holding it until death. At his command post, Horrocks felt an admiration for these stubborn parachute and panzer fighting men who, with chaos and dissillusionment on all sides, and with their numbers reduced by the bitter fighting west of the Rhine during the past six weeks, refused to budge until killed or rooted out of the rubble by British bayonets and grenades.

Late that morning, Horrocks received a message that stunned him: Major General Thomas Rennie, the dashing leader of the 51st Highland Division, had been killed by a mortar shell while up front. Horrocks reflected that Rennie had had a premonition of his death. In all previous operations, Rennie had carried out his role cheerfully, but in preparing for the Rhine crossing Rennie had found fault with every phase of the plan.

Rennie's death could not have come at a worse time—all three of his brigades were deeply engaged in heavy fighting. General Horrocks was rushed across the Rhine by Buffalo and summoned the three brigade commanders. He found them depressed and shaken by the death of their popular leader. Horrocks promptly appointed Brigadier James Oliver, commander of the 154th Brigade, to replace Rennie, with an exhortation to "vigorously press the attack."

Soon Horrocks's corps was struck with a violent counterattack by the 15th Panzer Grenadier Division. The situation became highly confused and the battle lines were fluid. One platoon of the 1st Black Watch Regiment could not be located. Nineteen-year-old Lieutenant J. R. Henderson volunteered to lead a patrol to locate the missing unit.

Henderson had gone only a short distance when he and his patrol ran into heavy German machine-gun fire. Ordering his patrol to take cover, the lieutenant and one man with a Bren gun went forward. Suddenly an enemy automatic weapon opened up at point-blank range. The fusillade killed the Bren gunner and knocked the revolver from Henderson's hand. Undaunted, the lieutenant charged the machine-gun nest alone and hacked the German gunner to death with his entrenching shovel.

22

The C-46 Deathtraps

Hard on the heels of Edson Raff's combat team, 72 of the new C-46 transport planes carrying Colonel James W. Coutts's 513th Parachute Infantry Regiment and its attached 466th Parachute Field Artillery Battalion were flying in a clear sky toward the Rhine River. The C-46 was considered to be a vast improvement, with two doors, one on each side, permitting paratroopers to bail out twice as quickly as from the single-door C-47s. Unknown to Coutts and his men, the C-46s, in combat for the first time, would prove to be fiery death traps.

Colonel Coutts had always been a scrapper. As a welterweight with a knockout punch in either hand on the West Point boxing team, he had won most of his fights. He had been a company commander in the 501st Parachute Infantry Battalion, America's first unit of that size, and he and his 513th Parachute Infantry Regiment had fought with distinction at Flamierge in one of the bloodiest actions during the Battle of the Bulge.

In one C-46, Private First Class John Magill, the forward observer from Sugarloaf, Pennsylvania, looked down the aisle at the teenage replacement whom Magill had taken under his wing. The boy was seeing his first action; his features were fixed in a haunting stare and he looked green in the face. Magill thought, I wonder if I look the same way? His stomach was churning too as his first combat jump was only minutes away.

Platoon Sergeant John K. Doyle was waddling down the narrow aisle of another C-46 carrying a box of cheese and peanut butter sandwiches. Someone had handed the box to him just before takeoff with the reminder, "You fellows might get hungry on the ride." In fact, food was the furthest thought from the troopers' minds. "Take one," Doyle suggested. "You might not get to eat for a long time." There were no takers. The sergeant hoped that as he held out the sandwiches his hand did not betray the nervousness he felt.

Returning to his bucket seat, Doyle thought he should practice what he preached. He bit into a peanut butter sandwich, but his mouth felt as though it had been filled with drying plaster. He coughed and gagged, taking out his canteen to wash down the peanut butter with water. In other circumstances, the platoon sergeant's discomfiture would have brought gales of laughter. Now the men only stared dully at Doyle.

Minutes later, Sergeant Doyle's C-46 reached the Rhine. A call rang out: "Stand up and hook up!" The men struggled to their feet. Doyle bent down to peer out a window and could see thick puffs of black smoke all around where antiaircraft shells were exploding. "Shrapnel!" someone called out. The grim-faced troopers fought off an overpowering urge to leap out of the aircraft, to get away from the death-dealing explosives.

Colonel Coutts, known to friends as Lou, was standing in the door of his C-46 after the red light flashed on. Ready to bail out right behind him was the famed *Life* magazine photographer Robert Capa.[1] As the aircraft passed over the Rhine, Coutts heard several loud cracks nearby. Moments later he glanced back and saw blood running down the walkway on the floor. He knew that one or more men had been hit.

Next to the door where Coutts was standing, shrapnel had ripped a hole in the fuselage. He instinctively reached out to it and saw that the hole was larger than his hand. One of the plane's motors burst nto flame. Now a sense of choking panic gripped the troopers. A sergeant, not knowing the motor was on fire, unhooked the harness of a wounded man stretched out on the floor.

Coutts knew the C-46 was doomed. The flames, fanned by the wind, were now licking at the wing. Trying to keep his voice calm, the colonel shouted to the sergeant, "Hook him up again and push him out the door!" The static line would open the semiconscious trooper's parachute, but whether he would survive a jolting landing in which he could not brace himself to absorb some of the shock was doubtful. The alternative, however, was for the wounded man to remain in the flaming C-46, which was clearly going down. Two men lugged the bleeding trooper to the door and pushed him through the opening. Few expected him to survive.[2]

The C-46 was down to 542 feet. The green light flashed on and Coutts and his stick rapidly bailed out the two doors. It was the lowest jump the colonel and his regiment had ever made, something Coutts was vaguely aware of in the few seconds before he crashed to the ground in a hail of bullets.

The colonel removed his chute, found his radio operator, and told him to try to open communications with the three parachute battalions. Nearby a young lieutenant was lying on the ground, struggling to get out of his harness.

"What's your problem?" Coutts asked.

"I guess I'm nervous or excited." The young officer had forgotten to take the safety pin out of his quick-release apparatus, which the colonel did for him.

"Now go get your platoon together," Coutts said.

Nearby, *Life* photographer Bob Capa was also having difficulty getting out of his parachute harness. The naturalized American was pinned to the ground by bullets whistling overhead and struggling with his quick-release mechanism. In his frustration he was cursing loudly in his native Hungarian. "Hey, buddy," a nearby prone trooper called out to Capa, "them Jewish prayers won't do you any good here in Germany."

Studying his maps and terrain features, Colonel Coutts quickly realized that his regiment had been dropped off target and was partially in the zone of the British 6th Airborne Division, so he contacted his battalion commanders and told them to head south to the designated DZ, then to swing east and seize the bridges over the Issel River.

Coutts, with his Tommy gun clutched in one hand and ready for instant action, marched off after picking up a trooper who was lugging one of the new 57-millimeter recoilless rifles, a sort of hand-held artillery piece. The colonel's concern was that his lightly armed paratroopers would run into German tanks, a concern that soon proved valid. He spotted German soldiers hurriedly removing camouflage nets from five or six panzers.

Coutts shouted for the trooper with the recoilless rifle. "Get behind that hedge," he ordered, "and take a couple of shots at those Kraut tanks."

"But colonel, we've got to know the range," the parachutist said.

"Try two hundred and fifty yards."

Coutts loaded the trooper's tube, stepped aside to avoid the rocket exhaust, and saw the shell streaking out of the muzzle. The colonel had only hoped to cause the enemy tankers to take cover if the round hit near them. But the shell crashed into one tank, which apparently had ammunition stacked on its deck, and the panzer erupted in a tremendous roar. Flames shot into the sky. Unnerved by the awesome explosion, the other German tanks spun around and fled eastward. Coutts let out a sigh of relief and clapped the gunner on the back.

Earlier, in the run-in to the 513th Parachute Infantry's drop zones, other C-46s were being shot up. The plane carrying Lieutenant Colonel A. C. Miller, commander of the 2nd Battalion, was flying so low that rifle and machine-gun fire were coming through the floor. A number of men were hit, and the walkway was flowing with blood. An excited crewman dashed into the cabin and shouted, "The copilot's been hit!" Moments later Colonel Miller and his men bailed out into a torrent of small-arms fire, which raked them all the way to the ground and after they landed. Several troopers were killed on the way down, their heads bobbing grotesquely and blood gushing from their bodies. The sky was filled with burning and smoking C-46s.

One of these fiery aircraft was carrying the 513th regimental surgeon, Major William W. Moir from Stillwater, Minnesota. A veteran f previous combat jumps, Moir had bailed out in North Africa with the 509th Parachute Infantry Battalion in November 1942 and had won the Distinguished Service Cross there, the first American paratroop officer to receive that award.

Just after his C-46 crossed the Rhine, Moir felt the plane being jolted twice in succession as shell fragments struck the aircraft, and moments later a motor broke out in flames. Those aboard knew the plane was doomed, but the pilot shouted back, "I'll try to hold it on course until we reach the DZ!" With smoke billowing into the cabin, the green light flashed on. Major Moir tossed an equipment bundle out the door, then bailed out. Moments after the stick had cleared the cabin, the pilot tried to make a crash-landing to save the C-46. Both he and the co-pilot were killed, but the crew chief survived, although seriously injured.

Reaching the ground, Moir scrambled into a nearby bomb crater as the terrain was being swept by machine-gun bursts. A short time later he was joined in the hole by the regiment's dental officer, Captain "Odo" Odorizzi of Hurley, Wisconsin. Seconds later the two men looked up to see a descending parachutist about to land nearby. The trooper's clothing was on fire, apparently ignited by incendiary bullets, and he was screaming in agony.

Once the man landed, his clothing still smoking, Major Moir and Captain Odorizzi left the relative safety of the bomb crater and ran to the aid of the severely burned paratrooper. Extinguishing the flames, Moir and Odorizzi lugged the man to a ditch, where the regimental surgeon began treating him. Moir, intent on his task, was only vaguely aware that two other figures had leaped into the ditch directly behind him. The surgeon instinctively glanced over his shoulder and felt a surge of alarm—the newcomers were armed German soldiers. Moir and Odorizzi had no weapons.

The two American medical officers were startled by the sudden confrontation—but so were the Germans. For several moments the adversaries crouched motionless and stared at each other. Finally, Major Moir broke the impasse. Speaking in passable German, he said, "Our paratroopers are all around you. You haven't got a chance to get out alive. You'd do better to surrender to us than be killed."

While the two Germans were hardly surrounded in such a fluid situation, Moir's viewpoint apparently made sense to them. They pitched their weapons to the ground and tossed away their helmets.

As the C-46 carrying Lieutenant Maurice Leland, machine-gun platoon leader in the 3rd Battalion of the 513th Regiment, passed over the Rhine it was struck by an antiaircraft shell. The door through which Leland and half of the other troopers had intended to jump was engulfed in flame, so they had to wait until the opposite opening was cleared before they could jump.

Nearing the ground, Leland saw he was going to land in a farmyard full of big trees. He saw a group of German soldiers running for the farmhouse just as his parachute caught in the top of a tall tree. Flicking the new quick-release harness, Leland plunged 20 feet through branches that inflicted numerous cuts on his face and hands. Thankful to be on solid ground but aware that there were armed Germans in the farmhouse less than 100 feet away, Lieutenant Leland felt a surge of alarm—his rifle was in an equipment bag attached to his harness high in the tree.

He dashed behind a nearby outhouse, feeling foolish with only a trench knife for a weapon. Flashing through his mind were the constant warnings of combat instructors in training camps—your rifle will be your best friend, don't ever part with it on the battlefield. If he tried to climb the tree to get it now, he most certainly would be spotted and shot.

Eventually, Leland slipped from his cover and obtained a rifle from a wounded paratrooper. He located several of his comrades and led the little group back to the farmhouse where the Germans were holed up. After a brief firefight the enemy inside surrendered. Holding their arms overhead, about 15 Feldgrau filed out.

Elsewhere, Private Charles L. Worrilow, a 23-year-old rifleman from Lima, Ohio, and other members of his 513th Regiment stick landed far from the DZ—nearly six miles away. As soon as he touched down in the center of a large field, Worrilow's first act was to satisfy an overpowering need to urinate. While thus engaged he glanced off into the distance and saw three gliders swooping in for a landing. He tensed in horror as the gliders smashed into power lines. The motorless, flimsy craft spun down from the wires, flinging hapless glidermen out of the fuselage. Worrilow saw one glider fighter land in a tree, where he remained motionless in the grip of its branches. There was nothing the paratrooper could do for the victims. He buttoned his jumpsuit and headed for the DZ.

Private First Class Lynn Vaughn of the 513th parachuted into a tree. Underneath were three confused Germans. Vaughn quickly shucked his chute and slid down the tree. He shot one German on the way down and the other two surrendered. Sergeant Curtis Gadd of Cleveland, Ohio, while descending from a burning C-46, spotted a German soldier frantically beating a horse into a gallop across a field. Gadd unslung his Garand rifle in midair and shot the rider.

The executive officer of Lou Coutts's 513th Regiment, Lieutenant Colonel Ward Ryan of Fort Atkinson, Wisconsin, dropped with his entire stick directly onto a German command post guarded by about a platoon of soldiers. Several of Ryan's stick were shot before they reached the ground, but the regimental exec and his troopers battled the Germans hand to hand until the enemy position was wiped out.

Private First Class Harry Boyle of the 466th Parachute Field Artillery Battalion would be last man out of his C-46. Standing and ready to bail out, he and his comrades heard ripping sounds as small-arms fire from the ground pierced the skin of the aircraft. Boyle, like the others, was ready to leap. The trip had been the most turbulent he had ever encountered; the man opposite him had vomited several times during the two-and-a-half-hour flight to the DZ outside Wesel.

Shortly after Boyle's chute popped open he became aware that streams of tracers were zipping past him, so he feigned being killed and hung limply in his harness until touching the ground. Small-arms fire was raking the DZ. It seemed to be coming from all directions. Boyle slithered into a gully about 100 feet away and began looking for a target. He spotted what looked like a German helmet protruding above a ditch and drew a bead on it. When Boyle was ready to squeeze the trigger, the target shifted and he saw that the figure was wearing a chartreuse scarf—which each artilleryman in the 466th had been issued for identification.

"That trooper will never know that a scarf kept him from getting a bullet through his head," Boyle reflected.

Slipping out of the gully, Boyle went in search of comrades. Fierce firefights were raging all around him. He came upon Captain Charles H. Jones of his artillery battalion, who was limping along at the head of a small band of troopers. Jones had been shot through the leg just above the ankle, the bullet piercing one side of his jump boot and coming out the other side. He was ranting at the Germans for ruining his prized footwear, coveted symbol of the American paratrooper. A medic appealed to Captain Jones to drop out and have his painful bullet wound treated. "Hell, no!" was the reply. Jones and his men continued on their way.[3]

Moving onward, Boyle reached a farmhouse where a paratrooper was seated on the ground still strapped in his harness, his parachute canopy hanging from the top of a tree. The trooper was in agony, having been shot through the thigh. Stretched out on the ground 25 feet away was an elderly German farmer, quite dead, with a bullet hole in his head. The wounded parachutist was weeping in frustration.

"I got hung up in this goddamned tree," the trooper gasped, "and while I was struggling to get out of the harness this old bastard rushed out of the house and shot me. Then I plugged him."

As Boyle had feared, the entire 466th Parachute Field Artillery Battalion had dropped onto a hornets' nest. Nine over-anxious troopers had jumped west of the Rhine, but 376 artillerymen and 12 howitzers had landed precisely on the designated DZ. The area was thick with German soldiers, and fierce firefights broke out immediately as the artillerymen fought desperately to clear their drop zone. In one battery, all the officers were killed or wounded within a few minutes after they hit the ground.

When Lieutenant Colonel Kenneth L. Booth of Fort Smith, Arkansas, commander of the artillery battalion, slipped his parachute to a landing in a bullet-swept orchard, he thought that a house only 200 feet away looked very familiar. He was puzzled. Why should a German farmhouse east of the Rhine appear to be one that he had seen before? Then he knew the answer: for days prior to the jump he had studied photos of this house, which he had picked for his command post. But never in his wildest dreams did he think he would drop a stone's throw from it. Colonel Booth and other troopers ran a number of Germans out of the house and set up the artillery battalion CP.

Private First Class John Magill, the artillery forward observer, and two companions were moving toward his infantry company's objective. With him were Captain Warren P. Elmer and the teenage radio operator, seeing his first combat, whom Magill had taken under his wing in recent days in order to sooth the boy's intense fears. Suddenly the three troopers were raked by machine-gun fire. They flopped to the ground briefly, then began leap-frogging forward in a series of short rushes. During one rush, Magill felt a heavy blow on his back, which knocked him to the ground. A bullet had ricocheted off the radio he was carrying and struck the teenager, killing him instantly. The boy had tasted battle for less than an hour.

When the artillerymen, fighting doggedly as infantrymen, had cleared their DZ, they had killed some 50 Germans, taken 320 prisoners, and captured ten 76-millimeter guns, eight 20-millimeter guns, and 18 machine guns. By 1:00 P.M. Colonel Booth had his howitzers in position to fire, and, while Colonel Lou Coutts and his men were fighting their way southward toward the 513th Regiment DZ, the howitzers fired at several German strong points holding up Coutts's infantrymen. It was a unique situation. Booth's artillerymen were firing from *behind* the Germans and *toward* Coutts's advancing force.

"What would the artillery school at Fort Sill have to say about this?" Colonel Booth mused to an aide.

Meanwhile, at an airfield at Achiet, France, remnants of the 313th Troop Carrier Group, which had carried Lou Coutts and his men into battle in the new C-46s, began to trickle in. It was immediately evident to anxiously awaiting ground crews that the troop carrier group had suffered a disaster. German antiaircraft guns had taken a toll of 19 C-46s shot down or fit only for salvage, and another 38 damaged, many of them severely. Of the 19 planes lost, 14 had gone down in flames. Angry survivors claimed that the C-46s seemed to catch fire each time they were hit in a vital spot.

Early in the afternoon on the airhead, Company E of the 513th

Parachute Infantry was attacking along a railroad track and had reached a point 250 yards from a large, sturdy farmhouse. The structure was being used as an enemy headquarters and defended by a large force of Germans with machine guns, rifles, and four artillery pieces. The enemy contingent was full of fight. One parachute platoon assaulted the house and was pinned down by heavy machine-gun fire after advancing only 50 yards. Lying dead at the head of the exposed platoon were its lieutenant and platoon sergeant.

Private First Class Stuart S. Stryker of Portland, Oregon, was in a relatively safe position at the rear of E Company when he saw that the exposed platoon was at the mercy of the enemy machine-gunners. Armed only with a carbine (which troopers called a "junior model" rifle), Stryker dashed to the head of his unit and, standing upright in full view of the Germans, brandished his weapon and shouted, "Follow me! Let's go get 'em!"

Inspired by Stryker's exhortation, and despite the machine-gun bullets lacing the air all around them, other troopers scrambled to their feet and charged toward the German-held house. The enemy fire increased in intensity, and Stryker was cut down only 25 yards from the structure. As he lay on the ground, the troopers he had led in the frantic dash stormed forward and surrounded the strong point.

After a fierce firefight, white flags appeared in the windows of the house, and E Company took over 200 prisoners and freed three crewmen of an American bomber who were being held in the basement. Stuart Stryker did not live to see the enemy surrender.

Shortly after 1:00 P.M. Private First Class Harry Boyle of the 466th Field Artillery Battalion heard a tremendous roar just over a line of trees, a noise so loud that it drowned out the heavy crescendo of small-arms fire. Boyle looked up to see four B-24 four-motored bombers zooming in only 400 feet above the ground. As he and his comrades gawked, waved, and cheered, bundles of ammunition and medical supplies began dropping out of bomb-bay doors, swaying back and forth under parachutes. Many German gunners turned their full attention to these low-flying, lumbering aircraft, and Boyle was dismayed to see one of the B-24s burst into flame and crash.

"If I ever get out of this mess," Boyle declared solemnly to a comrade, "I'm going to tell each bomber and troop carrier pilot I see that they've got a lot of guts!" Others nodded in agreement.

By 2:00 P.M. Colonel Lou Coutts reported to the division commander, General Miley, that all objectives had been seized. The 513th Parachute Infantry drop zones had been cleared, the Issel River bridges taken intact with the help of General Bols's British Red Devils, and 1,100 German prisoners were in tow. Coutts was in his CP when he looked up to see a haggard figure with a used parachute slung over his shoulder. The young air corps officer looked tired but cheerful. The colonel thought he recognized him, and he did. It was the co-pilot of the burning C-46 Coutts and his stick had jumped out of that morning.

"My God, I'm glad to see you!" Coutts blurted out. He quickly added, "But I sure as hell didn't expect to! How did you do it?"

"Well, when you guys jumped out," the boyish-looking lieutenant began, ignoring the colonel's eagles on Coutts's collar, "the C-46 began floating up from 542 feet. Our pilot knew where his friends were, so he gradually gained altitude, circled back over the DZ, and headed the plane toward Belgium. He set it on automatic pilot and the crew chief, he, and I all bailed out safely while the plane was still burning."

"Then what in God's name are you lugging that damned heavy parachute around with you for?"

"Colonel, in the air corps if you parachute out of a plane and don't bring the chute back you get your paycheck docked thirty-two dollars and eighty-four cents."

Coutts and other troopers in the CP doubled up with laughter. The colonel gave him a receipt for his parachute, "so you won't have to drag it around with you for days." Coutts also gave the pilot something else—a carbine and directions to a nearby foxhole, where he was posted as a security guard for the CP.

Several hundred miles from the flaming Rhine, one of General Miley's paratroopers, 20-year-old Private First Class Bart Hagerman, was listening to popular music over BBC radio. Hagerman had received a disabling wound during the Battle of the Bulge eight weeks previously, and was now lying in a hospital bed in Scotland waiting for an airplane to return him to the States for treatment. His thoughts were of his home in Bowling Green, Kentucky, and he wondered whether he would be able to visit there soon.

Suddenly the music halted. An announcer broke in with a bulletin: "The U.S. 17th and British 6th Airborne Divisions this morning landed east of the Rhine River near Wesel, Germany, in the largest single airborne assault in history. Heavy fighting is raging."

Hagerman lay silently for several moments. He had had no idea his outfit was to make an airborne landing. Then he broke down in tears, anguished at the thought of not being able to fight alongside his comrades.

23

Glider
Fighters
Pounce

Earlier that morning of D-day for Varsity, Colonel Adriel N. Williams of Shelbyville, Kentucky, commander of the 436th Troop Carrier Group, had been piloting a C-47 towing a glider and heading for the landing zone of the 194th Glider Infantry Regiment of Miley's 17th Airborne Division. The LZ was a rectangle of ground more than two miles long and a mile wide, about two miles northeast of Wesel. In the greater scheme of things, the landing zone was merely a dot on the landscape, and it would take all the skill that the glider pilots could muster—plus luck and divine intervention—for hundreds of the pilotless aircraft to crash-land in this relatively confined space without smashing into each other.

Colonel Williams was clutching the controls tightly and, like all other pilots in the tug-glider stream, was having a difficult time keeping in formation. "Worst turbulence I've ever experienced," the veteran flyer remarked to his co-pilot. So strenuous was the task of holding the tug-glider tandems in formation that the pilots and co-pilots had to alternate at the controls in 15-minute stints.

Commanded by Colonel James R. Pierce, the 194th Glider Infantry combat team was winging to battle in 906 Wacos pulled by 610 C-47s of the 435th, 436th, 437th, and 439th Troop Carrier Groups. Along with the infantry, they flew the 680th and 681st Glider Field Artillery Battalions, the 139th Airborne Engineer Battalion, and the 155th Antiaircraft Battalion. In the Wacos were 3,594 glidermen and 654 tons of cargo, including 208 jeeps, 101 trailers, and 84 mortars and guns. On board many gliders were large amounts of ammunition which, the men riding in them were fully aware, would result in the flimsy crafts disintegrating in flames if they were struck by anti-aircraft fire.

Crammed into a Waco glider seat with 12 grim-faced comrades was Private First Class Eugene Herrmann, a 21-year-old mortarman who, like others in the long sky train, kept conjuring up pictures he had seen of smashed and splintered gliders in Normandy and Holland. He tried to banish such thoughts from his mind—but the vision kept returning. Herrmann regularly glanced to one side, where another motorless craft was virtually brushing his wingtip. If the two gliders locked wings, the two crafts would plummet to the ground.

The Waco's wings shook and shimmied. Herrmann had never seen them vibrate so violently. At one point, without warning, the glider dropped suddenly, shaking off helmets and causing some men to wind up on their knees in the aisle. One man bashed his head against a metal framework. Each knew that if the glider shook hard enough it would break up. As the slipstream whipped the cloth-covered ribs of the craft, disturbing noises erupted. Nerves tingled. Was the glider breaking up?

Suddenly, the glider pilot called out, "It won't be long now." In the hazy distance Herrmann could see thick black smoke—the LZ. Below them he glimpsed the majestic Rhine, and seconds later the glider was cut loose and began a rapid descent. Everyone had weapons cocked and across their laps—they might have to scramble out with guns blazing.

Loud cracks rocked the glider—ack-ack, ominous and lethal. Everyone tried to hunch down lower except the pilots, who were desperately searching for the LZ. The flak grew heavier, and Herrmann's thoughts flashed back to the previous night's broadcast warning by the Berlin Bitch: *You boys won't have to worry about getting your feet wet crossing the Rhine. Our flak will be so heavy you can walk down on it . . .*

Now Herrmann was aware that the landscape was flashing past. At 80 miles per hour the glider pancaked onto the ground and, skidding on its belly, smashed through a fence, hurdled two ditches, and kept going, leaving a trail of plowed turf behind it. Creaks and tears rent the air as the flimsy craft buckled under the shock. Finally, 150 yards from touchdown, it slid to a halt.

All on board relaxed momentarily and breathed sighs of relief. "Well, we made it," Herrmann remarked to his friend, Private First Class Phil Snow.

"Get the hell out!" someone shouted. No inducement was needed; bullets began ripping through the canvas skin. There was a mad scramble and in seconds the glider was empty. Herrmann glanced back to see the two pilots calmly gathering up their gear.

Some distance behind Colonel Williams's lead aircraft, Lieutenant D. F. "Dusty" Rhoades of Maryland, Wisconsin, checked his watch on spotting the Rhine River in the distance. He grinned and remarked to his co-pilot, Lieutenant C. W. Alderdyce of Toledo, Ohio, "Right on the button!" Pilots on a mission always liked to take off and arrive on schedule. It was a good omen.

Alderdyce started to reply when both men saw a C-47 towing two gliders a few hundred yards to their front burst into flames. The wounded aircraft could be seen to shudder and wobble as it burned, but the pilot held it on course until he released both Wacos. Just as Rhoades's C-47 arrived at the glider-release point, a burning plane plunged by and headed for the ground. No one said anything as the grim-faced Alderdyce reached up to pull the handle releasing his two gliders. On all sides of Rhoades's and Alderdyce's tow plane were puffs of black ack-ack smoke.

Suddenly freed from its heavy burden, the C-47 surged forward, pushing the crew members back hard against their seats. Rhoades banked the craft to return home when radioman Staff Sergeant Finous

L. Rood from Dallas, Texas, called out, "Oh, oh. We got it good that time!"

Rood leaped to his feet, looked out the window at the wing, then crossed to the other side and stared intently again. The others looked curiously at the radioman. They had felt nothing and thought Rood might have come down with battle jitters. "Damn it," he said, "I know we were hit, but I don't see anything. I felt it in my feet." Only when the C-47 reached base and tried to land would Sergeant Rood be proven right: the hydraulic system had been hit and the wheels had to be let down manually, the flaps came down but couldn't be controlled, and the brakes were stuck despite priming with emergency fluid.[1]

Elsewhere in the tug-glider flight stream, Captain Victor Deer and his co-pilot, Lieutenant Eldred Tracha, were intently scanning the terrain for their LZ after crossing the Rhine. To keep on course, they flew directly into a barrage of black smoke puffs. Suddenly the ship was rocked when a shell exploded in the passageway. Shrapnel ripped into the navigator, Lieutenant Bryce Nelson, and Sergeant Carroll, the radio operator. Deer's C-47 began to career crazily, so the glider pilots in tow cut loose from the tug.

Technical Sergeant Brack LeFevre, the crew chief, had escaped the shrapnel as he had been standing in the dome with Aldis light in hand to flash a signal to the gliders when it was time to cast off. LeFevre rushed to the aid of his two seriously injured comrades, but each was unconscious and bleeding profusely. Again the C-47 was jolted—another shell had struck the doomed craft in the wing and set the gas tank ablaze.

LeFevre called out to Captain Deer that the two crewmen were still alive but could not jump. Deer made an instant decision: in an effort to save Nelson's and Carroll's lives, he would risk his own by making a crash landing with the hope of getting medical attention for the grievously wounded fliers. All knew that his chances of bringing the crippled aircraft down safely were minimal.

There was no time to try to get back over the Rhine, so Deer began frantically searching for a suitable field on which to crash-land. He called to LeFevre, "Bail out while we've still got enough altitude!"

As Sergeant LeFevre turned and headed for the fuselage door, he heard Captain Deer call to his co-pilot, "Eldred, get the hell on out the escape hatch!"

"Like hell I will!" LeFevre heard Tracha yell back. "If you're riding it on down, I'm going with you!"

LeFevre felt his parachute to make certain it was in place, leaped through the open door, pulled the rip-cord, and seconds later struck the ground with terrific impact. He had jumped from such a low altitude that his parachute barely had time to open. As he lay on the ground in his harness, his head spinning crazily, he could see the C-47 carrying Captain Deer and his other comrades flying in the distance. It was now a solid sheet of flame. Moments later the aircraft exploded in midair. LeFevre buried his face in his hands and wept.[2]

The hazy morning sky east of the Rhine was dense with hundreds of gliders descending after cutting loose from their C-47 tugs. Machine-gun tracers crisscrossed the air and puffs of black ack-ack smoke erupted in the wake of explosives which sought out the flimsy, helpless Wacos. Two-thirds of the gliders carrying Colonel Jim Pierce's 194th Glider Infantry combat team were struck by the torrent of German fire, but most soared onward. On all sides, dust plumes spiraled into the sky as the Wacos skidded on their bellies over lengthy stretches of rough terrain, the grim-faced riders holding on for dear life and praying that a tree or a pole or a stone wall would not suddenly loom in their path.

Some Wacos did smash into obstacles and crumple like accordions. Screams pierced the din of gunfire as glidermen were slashed and pinned in the twisted wreckage, many with hideously broken bones. Other crackups resulted in all riders being killed instantly. A glider in which Private Lew Morris was a rider hit a high-tension wire and flipped over, landing upside down. Morris's carbine stock was shattered and he was knocked semiconscious. A comrade's teeth were smashed out and blood poured from his mouth. A young lieutenant had his foot severed, and it was lying in the wreckage.

The LZ was infested with entrenched German riflemen and machine-gunners. Each building seemed to have a sniper or a squad pouring fire out of windows. There were several batteries of 20-millimeter flak guns, five or six artillery pieces, and numerous mortars on the LZ, most of them in action.

Wild confusion reigned. More than 150 battles raged at various points. Grand strategy was out the window. Glidermen struggled face-to-face with the enemy using grenades, rifles, pistols, and Tommy guns. Part of Company F charged a German command post with such swiftness and ferocity that a colonel and all of his staff were seized. As the German was being escorted out the door with a Tommy gun in his back, an aide, bewildered by the sudden onslaught from the sky, rushed up brandishing a briefcase and called, "*Herr Oberst* [Colonel], you've forgotten your maps!"

The blunder proved to be a bonanza. The maps contained overlays of all German defenses in the area.

Moments after clambering out of a smashed Waco, Simon Sculkowski, John Marcyyk, and Joe Dubskey spotted 10 or 12 fleeing Germans about 100 yards away. Marcyyk, excited on seeing his first armed enemy, took out his Colt .45 pistol and emptied the clip in the direction of the departing Germans, the bullets of the short-range weapon plowing into the ground only 70 feet away. His frantic efforts drew chuckles from his two comrades despite their tense situation.

Minutes after landing, men of the 139th Airborne Engineer Battalion were involved in savage firefights with Feldgrau barricaded in houses. One platoon had wiped out defenders in two structures, then approached a third. Private Eric Shinn, who had been seriously wounded but had crawled away after being hit, told the platoon leader that this house was a strong point in the German defenses. They also learned that Private Dowlan, a comrade, was still lying seriously wounded near the house. Lieutenant Goodman asked for two volunteers to join him in an effort to rescue the wounded gliderman from under the Germans' noses. Two privates agreed to go along. As Goodman and his two men neared the house, there was a sudden chatter of a machine gun. Two of the glidermen were killed, the other seriously wounded. Angered on seeing their comrades gunned down, the remainder of the platoon stormed the strong point, killed and wounded a number of Germans, and took 30 prisoners.

On landing, a glider carrying medics and medical supplies was hit by mortar shells, and the Waco and the supplies went up in smoke. One medic burned with it. Staff Sergeant Totten, who had scrambled out moments before the Waco burst into flame, knelt under intense enemy machine-gun fire and tended to several wounded. He heard a seriously hurt man pleading for help some distance away and dashed to his comrade's aid. A German sniper took potshots at Totten every foot of the way.

Medic Totten tried to drag the wounded gliderman to cover, and while so doing the injured soldier was hit twice more by bullets and Totten himself was shot. Totten, despite his intense pain, finished moving the other man to relative safety and gave him first aid. The medical sergeant then hobbled back into the fire-swept field to attend to other wounded glidermen.

As Corporal Thornton E. Hiese of the 139th Airborne Engineers dashed out of his glider, a bullet knocked his rifle from his hands just as five Germans were emerging from the basement of a farmhouse. Weaponless, Hiese tackled the first German, who was brandishing a pistol, knocked the man to the ground, and wrested the Luger from his hand. Bewildered by the sudden one-man onslaught, the other four enemy soldiers surrendered just as Hiese was ready to unload the Luger at them.

A Waco carrying a nine-man squad of engineers landed three miles off target, directly in the center of a German 88-millimeter battery. The enemy artillerymen raked the Americans as they rushed out of the glider, killing one man and seriously wounding two others. In the confusion, two more in the squad simply vanished. The remaining members of the squad battled the German gunners for two hours, killed or wounded a large number of the enemy, and took the survivors prisoner. The airborne engineer squad then spiked the gun barrels with thermite grenades, took over a German staff car, and herded their prisoners before them back to their own battalion command post.

Major Charles "Chuck" Williams of the 437th Troop Carrier Group was at the controls of his C-47 as it neared the Rhine River, a pair of Wacos soaring along gracefully behind it. Varsity was to be the final flight for the veteran C-47 named Redcliffe Able. It had survived the

D-Day assault in Normandy, flown in the invasion of southern France, dodged thick flak in the Holland operation, and weaved through lead-punctured skies over snow-blanketed Bastogne. After this mission it was to be put out to pasture.

Moments after the pair of towed Wacos were released, Redcliffe Able's luck ran out. In quick succession the C-47 was riddled by 8 to 10 hits from a 20-millimeter gun. The plane's belly, engine, left wing, and elevators were damaged. Major Williams heard thuds behind him and glanced back to see crew members bleeding. Radio operator Staff Sergeant James Lyons and crew chief Corporal Francis Gildersleeve had been peppered by white-hot flying metal.

With the C-47 shuddering violently, smoking, and chugging along on one motor, Williams and his co-pilot, Captain Joe McGloin, decided to try to limp to their home base. But as they passed over the Rhine, the surviving motor conked out. Switches were cut to reduce the chance of fire in a crash landing, and Major Williams glided toward a large field. As Redcliffe Able skidded along in a wheels-up landing, kicking up a plume of dust, there was a wrenching of fabric and metal. Parts damaged by flak began to tear loose. The crew braced as the C-47 smashed into a grove of trees and folded in accordion fashion.

Major Williams was jolted forward, his head striking the instrument panel, causing blood to gush from a deep gash. Captain McGloin's legs were badly crushed and splintered as the cockpit floor doubled up underneath him. The two wounded crewmen, Lyons and Gildersleeve, received new injuries when they were slashed about the head and body. Lieutenant Joe Salisbury, the navigator, was stunned when thrown violently against the instrument panel. In the back of each man's hazy mind was the terrifying thought that the demolished C-47 might explode at any moment. Despite his groggy condition, Salisbury managed to assist his injured comrades out of Redcliffe Able.

To the battered men of Redcliffe Able, it seemed a lifetime since the C-47 had been riddled by flak in the smoke-filled skies over Wesel. Actually, it had been six minutes.

Back on the LZ, Staff Sergeant Robert Thomson of Upsala, Minnesota, and some 30 of his fellow glidermen had assembled and were ready to move out in search of Germans. Ahead of them, a glider swooped down and skidded to a halt near the edge of a woods filled with birch trees—and Germans. As Thomson and his comrades looked on, a lone German dashed out of the woods into the clearing and pitched a grenade into the glider before the crew and passengers could emerge. The flimsy craft went up in flames as the enemy soldier dashed back for the trees. Thomson and his 30 comrades blazed away with Tommy guns and rifles at the lone German, but he apparently was leading a charmed life. He was swallowed up by the woods without being hit.

By noon, Colonel Jim Pierce's 194th Glider Infantry combat team was three-quarters assembled and German resistance was beginning to crumble. The Wehrmacht had based its hope for defeating the Allied airborne attack on knocking carrier and tow planes from the sky and on destroying or decimating the paratroopers and glider fighters as they were landing. The LZ was littered with the black skeletons of C-46s, C-47s, and gliders, but once it was evident to the Germans that the airborne invaders had landed in force, their will to resist began to wilt.

The flight of the British glider stream had been uneventful except for a tank falling through the bottom of one plane in midair, plunging the unsuspecting crew to their deaths. It was nearing 9:45 that morning when Brigadier G. K. Bourne was peering ahead at the pall of haze and smoke hovering over the landing zone of his British 6th Air Landing Brigade east of the Rhine. Bourne's glider-borne Red Devils were to land in a tight concentration around the heavily defended town of Hamminkeln, about five miles north of Wesel.

Flying at 2,500 feet with 440 large Horsa and Hamilcar gliders strung out for miles behind, Brigadier Bourne thought smoke-shrouded Wesel looked no different than the British cities of Manchester or Birmingham as seen from the air. The brigadier was inwardly cursing the orders he had helped write: each gliderman was to strap himself to his seat. Like every other Red Devil, Bourne wanted a better view of what was going on up ahead.

The brigadier, aware that the glider had been cut loose and was starting its descent, listened with intent interest to the two pilots, neither of whom had been on a combat mission before. "I can see the railway!" one pilot called out excitedly. Bourne felt a deep sense of relief; the railway was a landmark near the LZ. Minutes later the brigadier saw the landscape racing past his window, then felt the curious scraping sensation as the big glider's belly skidded along the ground and through a couple of fences before the craft jerked to a halt. Bourne jumped out and took cover behind an embankment. He heard heavy firing off in the distance, and the cracks of sniper fire around him. The brigadier was pleasantly surprised to determine that his glider had come to rest only 600 yards from the farmhouse he had chosen as his CP. Waving his men forward, he set out for the house and on arrival found a large number of German prisoners standing around docilely, having already been nabbed by other Red Devils.

Despite the shroud of smoke and haze and the ack-ack explosions dotting the sky, each Horsa or Hamilcar landed with almost pinpoint accuracy, each having been assigned a specific target. Two gliders carrying parties of the Royal Ulster Rifles and the Oxfordshire and Buckinghamshire Light Infantry landed on each side of a bridge over the Issel. Scrambling out of their gliders, the Red Devils pounced on the German defenders and in minutes had control of the river span.

When the American reporter for United Press, Robert Vermillion, got out of his parachute that morning, he began looking for his young British friend, Geoffrey Bocca of the London *Daily Express*, who had agreed to haul Vermillion's typewriter in his Horsa glider. He met a Red Devil officer who told him Bocca's glider had been smashed and the reporter killed. Vermillion was angry at himself for not jumping with his 25-pound portable Remington; now he would have to laboriously scratch out his eyewitness stories by pencil.

The American reporter had no way of knowing that Geoffrey Bocca was in fact alive, or of the ordeal Bocca had undergone that day. Bocca had been shot five times on landing, been captured, escaped, and been recaptured. One of his wounds came when he was slithering over the ground under heavy machine-gun fire in an effort to retrieve his friend Vermillion's typewriter.[3]

One of Brigadier Bourne's glider pilots, Sergeant J. H. Jenkins, and his co-pilot, Sergeant Andy Anderson, were almost giddy from the thrilling sight of the stream of tug-glider combinations, which stretched out to the front and back as far as the eye could see. Their flight from England had been almost enjoyable. Suddenly, over the Rhine River, their elation ceased: there were sharp cracks and black puffs of smoke. The Horsa shook and quivered. Moments later they spotted the green light from the Aldis lamp in the towing aircraft, and the glider was cut loose.

Sergeant Jenkins, a veteran of previous glider combat operations, fought off a slight surge of alarm—it was so smoky and hazy that he could not see the ground, much less locate a tiny landing zone. As the flimsy craft glided downward, pilots and Red Devils alike heard a frightening swishing sound—another glider, burning fiercely, had plummeted past them, barely missing their own glider.

Providentially, a hole opened in the thick pall and the pilots spotted the outline of a church steeple to their front. They let out whoops of exultation—the spire was their landmark. Soon Jenkins touched down in a plowed field and the glider bumped along for over 100 yards before lurching to a halt. Along the way one wing had been neatly clipped off by a tree trunk.

The Red Devils scrambled from the glider and were fired on as they dashed for cover. Jenkins and Anderson rapidly grabbed their Bren gun, leaped from the craft, and set up the weapon in a shallow trench. Moments later enemy machine-gun bursts were whistling past just overhead. When the shooting slackened, Sergeant Anderson remembered the thermos bottle of tea in the glider. Jenkins dashed to his nearby craft and returned with the container. Unscrewing the cap, the two men were downcast; a bullet had penetrated the thermos and all the tea had leaked out.

One British glider crashed on landing, and the two pilots were hurled through the glass covering the cockpit. The Red Devil passengers were seriously wounded or killed, and lay in grotesque postures within the twisted wreckage. The impact of the crash had bent the pilots' rifles into pretzel shapes. Groggy and disoriented, one pilot staggered back to the glider and instinctively recovered his bent rifle. He sat down on a piece of wreckage with his useless weapon across his knees and his head in his hands. He became aware of the sound of voices, looked up, and saw through hazy vision some 20 armed German soldiers standing in front of him. Alone, with his comrades dead or seriously injured and his rifle useless, the dazed pilot slowly realized that the Feldgrau were surrendering to him.

Not all Germans had become so demoralized. Captain Boucher-Giles, a British glider pilot, had made a relatively soft landing. But seconds later he and his comrades were peppered by machine-gun and mortar fire as they scrambled from the crate. Boucher-Giles had only six men with him, but all were armed with automatic weapons and they returned the fire until the enemy quieted down.

Boucher-Giles then prepared to unlock the tail unit in order to remove the jeep and trailer—but no one had thought to bring the key. The men hacked furiously with an axe at the wires holding the unit shut, but to no avail. The captain then wired the tail with explosives and was ready to ignite the charge when the tiny group was raked by machine-gun fire. Boucher-Giles felt an enormous jolt against his thigh and a searing pain surged through his body. He lost consciousness, but was dragged to relative safety in a ditch.

A Red Devil major was wounded, as was a private; another private was killed. Fortunately, several more British gliders landed nearby. Out of a Horsa stepped Nigel Brown, who by coincidence was the staff sergeant of Captain Boucher-Giles's troop. Brown was noted for his utter disdain for enemy fire. With bullets whizzing over the landscape, Sergeant Brown nonchalantly arranged his silk scarf. He simply would not think of going into action looking untidy.

Brown had participated in all the major glider operations, and, although always in the thick of the action, he had never been scratched by enemy bullets or shells. Comrades were understandably a bit superstitious about him, and he was much in demand as a foxhole partner due to his apparent immunity from harm. Along with Red Devils from the other Horsas that had landed nearby, Sergeant Brown attacked the German machine-gun positions and rapidly wiped them out. Later that morning, a Tiger tank shot the Bren gun out of Brown's hand, reducing the weapon to a smoking piece of twisted metal. Brown was not injured. He shrugged his shoulders, calmly picked up a rifle from a wounded comrade, and continued in the fight.

Casualties in the 6th Air Landing Brigade were considerable, but they would have been much heavier had it not been for the unforeseen *chance de guerre* whereby Colonel Lou Coutts's Americans of the 513th Parachute Infantry Regiment dropped in error onto a section of the British glider-landing area before the Horsas and Hamilcars arrived. Coutts, firing his Tommy gun at the head of his paratroopers, had immediately gone to work clearing the landscape of Germans without regard as to whose zone they were fighting in.

A veteran Red Devil officer who had seen Colonel Coutts and his men in action observed to a fellow officer: "I say, those American paratroop chaps have done a wonderful job. They are not just good fighters, they are very good."[4]

Even with the unexpected help of Lou Coutts and his men, of the 416 Horsa and Hamilcar gliders which reached the battlefield only 88 landed undamaged. The other gliders in the force had all been hit by flak or machine-gun fire, and 37 had burned to their airframes. Pilots of the Horsas and Hamilcars were particularly vulnerable. Between 20 and 30 percent of them were killed, wounded, or missing.

Shortly after 2:00 P.M. that day—less than five hours after the first paratroopers had jumped—General Bud Miley's Thunder from Heaven men made contact with the British 1st Commando Brigade, which had battled its way into Wesel the night before. At almost the same time, General Eric Bols's Red Devils linked up with the British 15th Division in Hamminkeln, six miles east of the Rhine.

24

"General, the German Is Whipped!"

Back at General Bill Simpson's Ninth Army press camp at Schloss Rheydt, Major Barney Oldfield, Simpson's press officer, had been saddled with new headaches in the ongoing war within a war—the continuing duel between American and British reporters covering Field Marshal Montgomery's 21st Army Group's Rhine crossings. Montgomery's press officer had set a 4:00 P.M. release date for any stories on Plunder/Varsity. Furious American newsmen were convinced that this had been a maneuver by Montgomery to assure that Fleet Street papers got first crack at one of the big stories of the war. Another Montgomery directive forbidding any mention of individual units was interpreted by American correspondents as a method for blacking out the participation of Simpson's Ninth Army and Miley's 17th Airborne Division.

At about the same time that American and British paratroopers were spilling out of troop carrier airplanes that morning, Major

Oldfield was sipping his fifth cup of coffee when a communications operator rushed up to him. "Major," the man called out excitedly, "our teletype has gone dead!" There was no power, he explained, and he could not raise anyone.

Curses rang out from several American reporters in the room. Major Oldfield quickly put in a frantic call to Ninth Army headquarters. He was startled by what he heard: *all* communications by land line to the rear from Ninth Army had mysteriously gone out. The radio link in Ninth Army still was operational, but on that link the messages went out in the open (uncoded), so that source was closely restricted.

"Maybe it's sabotage," suggested a signals officer on duty at Schloss Rheydt. "They could get hold of our line almost anywhere."

"Yeah, it's sabotage, all right," an American reporter exclaimed. "What's not known is which side did the sabotaging!"

Oldfield cautioned his men and the newsmen to remain mum while a check was run on the teletype cable. "Don't go jumping to conclusions," was the gist of his counsel.[1]

Meanwhile, in the late morning hours, Ninth Army war correspondents began to trickle in, mud-caked, red-eyed, near exhaustion. Some had made the nighttime crossing with assault troops. Richard C. Hottelet of CBS Radio and *Collier's* magazine stumbled into Schloss Rehydt with a heap of parachute silk thrown over his shoulder. A hot meal had been prepared for the eyewitness reporters, who knew that they could enjoy the rare treat, relax, unwind, and still have plenty of time to write their stories to meet the 4:00 P.M. release time. Just as Major Oldfield was preparing to chomp down on his first bite, one of his aides, a worried look on his face, pulled Oldfield away from the table.

"Captain Sam Brightman just called from Bradley's headquarters," the young lieutenant blurted out. "He said they'd been trying to raise us by teletype, but couldn't, so he gave me the message over the phone. SHAEF has overruled 21st Army Group's release time of four P.M. and the story can be released at twelve noon today."

Pandemonium erupted. Oldfield's announcement at the lunch table caused a great spewing of coffee. Chairs were shoved back and fell over with resounding crashes. Curses bounced off the ceiling. Not one of the correspondents who had risked his life to cover the Rhine crossings and airborne operation had written his story. There was loud and profane speculation as to who was trying to do what to whom at the higher levels of the Anglo-American command. Obviously, a great deal of jockeying for position was taking place in the rarefied atmosphere at SHAEF and at 21st Army Group.

Press officer Oldfield was in a cold sweat. He watched as the reporters—American and British—furiously pounded out their stories. The teletype was still dead. How would these eyewitness accounts get out? Oldfield felt a tugging at his sleeve and turned to look into the relieved, perspiring, and grease-covered face of his chief teletype operator. Out in the hall, the sergeant whispered, "The line's back in. We're ready to roll." Oldfield felt that the weight of a battleship anchor had been lifted from his shoulders.

"Ya know what knocked us out?" the operator added. "It was one of those dumb bastards with a bulldozer fixin' the road. He sunk his blade too deep at the shoulder and hacked through the Ninth Army cable with a hundred lines in it."

Five minutes later the correspondents began filing the stories that went out to the waiting world.

Nearly 4,000 miles west of the flaming Wesel caldron, the U.S. House of Representatives in Washington, D.C., rocked with cheers after Congressman Andrew J. May, chairman of the House Military Affairs Committee, told the news of the Rhine River crossings and the airborne assault at Wesel.

"The American armies have crossed the Rhine; the battle cry is 'On to Berlin!'" May shouted to thunderous applause.

House members promptly approved a resolution by Congressman John E. Rankin of Mississippi that "the Speaker send Lieutenant General George S. Patton, Jr., a congratulatory message on the magnificent showing he is making on the Western Front."[2]

No mention was made of General Courtney Hodges, whose First Army had stormed over the Rhine two weeks ahead of Patton, nor of General Bill Simpson, who crossed only hours after Patton had slipped over. For months, headline writers in Stateside newspapers had invariably referred to "Patton's army" or "Patton's tanks" or simply "Patton," while other armies were called by their numbers. So ingrained had the name "Patton" become in the home-front mind by these headlines that even Congress was equating any daring action along the western front with the colorful, dashing George Patton.

On the afternoon of D-day for Plunder/Varsity, Supreme Commander Dwight Eisenhower flew to Bradley's 12th Army Group headquarters at Namur. He had been up all night, but was beaming and talkative, and optimistic. Bradley advised the supreme commander that Hodges at Remagen and Patton at Oppenheim were "rarin' to go." Eisenhower gave his approval for them to break out of the two bridgeheads the following day.[3]

With the help of Bradley, Eisenhower drafted a signal for the Combined Chiefs of Staff, whose members at Malta had agreed on a strategic concept in which Montgomery would carry the ball in the north with Hodges and Patton in supporting roles. That concept had been steadily eroding in recent weeks, and now, with two "cheaply obtained" bridgeheads to go with Montgomery's crossing, had gone completely by the boards. Ike was, in effect, telling his bosses that he was now going to call the strategic tune.

In midafternoon of D-day, General Ridgway, commander of XVIII Airborne Corps, learned of the linkup of his two parachute and glider divisions with the ground forces. Ridgway crossed the Rhine with four aides in a Buffalo and set out on foot to find Bud Miley's 17th Airborne CP. Gripping his old bolt-action 1903 Springfield rifle in one hand, Ridgway led his party into some woods. Up ahead, the general heard a heavy thumping sound and motioned for all to take cover. Moments later a large, heavy-hooved farm horse came down the path with one of Miley's troopers astride the animal. A high black silk hat was tilted jauntily on the airborne man's head, and his Garand rifle was slung across his back.

Ridgway hopped out into the path in front of the dray horse, and the trooper reined in the snorting animal. Spotting Ridgway's two stars, the rider became confused, not knowing whether to hurriedly dismount and stand at attention, salute, remove his hat, or a combination of them all. The normally stern-featured Ridgway broke out in a belly laugh and the trooper, greatly relieved, joined in.

Meanwhile that afternoon, the 17th Airborne Division's surgical teams had been working in a hospital tent set up by glider-borne medics under incessant sniper fire only 55 minutes after the first glider had landed. In the first hours of airborne operations behind enemy lines there is no "front," so the hospital was operating with enemy soldiers on all sides.

Lieutenant Colonel Edward Sigerfoos, of San Antonio, Texas, surgeon of the Thunder from Heaven division, was ministering to wounded fighting men with his injured nose covered with adhesive tape, and limping painfully from patient to patient. Sigerfoos's glider had had its wing shot off and cart-wheeled to earth from a 50-foot altitude. As he crawled from the twisted wreckage, his first thought had been for the patient food the craft was carrying. To his dismay, it had been destroyed.

Much in the same condition as Sigerfoos was Major William Porter, medical inspector for the 17th Airborne, whose glider was shot up as soon as it lurched to a halt. He, too, was limping badly and his face was swathed in bandages.

One of the division's two surgical teams had been put out of action. One of its surgeons was killed, another missing, and a third had suffered a broken leg on landing. But by midafternoon major surgical operations were underway. Major Porter, meanwhile, had organized an impromptu commando party of jeep ambulance drivers and aidmen to try to fight their way through German positions and back to the Rhine beachhead for vitally needed food and supplies for the wounded. Porter had had to abandon the project after trying every road on the map and being forced back at all points.

Not far from the main hospital that afternoon, Private Edward J. Siergiej, a mortarman, was barreling along in a jeep stacked high with shells. His mortar section near the Issel River had nearly run out of ammunition and it had received word that German tanks and infantrymen were forming up for an attack. Siergiej had been ordered to take a jeep and try to find more shells.

"But I've never driven a jeep before," he told his platoon leader.

"Fine," was the reply. "This is a damned good opportunity to learn."

Siergiej secured the shells and headed back over an open field when his vehicle was spotted by Germans and mortar rounds began to explode around him. "If a shell hits this load of ammo, they'll sweep me up in a dustpan!" Siergiej thought. His comrades had been watching him approach their position with a great deal of glee and greeted him with: "Hell, you weren't in nearly as much danger from those Kraut shells as you were from your goddamned crazy driving!"

As the shadows began to lengthen, B. J. McQuaid, a correspondent for the *Chicago Daily News*, caught a ride with an ambulance which was moving rapidly through the Diersfordter Forest, the scene of some of the bloodiest fighting by both the 17th and 6th Airborne divisions. There was ample reason for the fast pace: the woods were reputedly full of snipers and isolated pockets of die-hard Germans. There was the chatter of small-arms fire, seemingly on all sides. The road was lined with grotesquely twisted corpses, most in Wehrmacht uniforms.

The small, neatly fenced fields were filled with the carcasses of cattle, horses, sheep, and goats, and with the bodies of men and the skeletons of shot-up, shot-down, cracked-up, and burned-up gliders. It was a scene out of Dante's Inferno, and it made McQuaid shudder.

One pattern was repeated over and over—a line of scorched glider skeletons, with bodies of four or five British or American boys crumpled on the ground where they had been gunned down as they emerged from the burning crates.

At a forward medical clearing station, McQuaid ran into Lieutenant Colonel Sigerfoos. Hundreds of casualties—American, British, and German—lay on stretchers in an open field. Those Red Devils who were not too seriously injured to talk, most of them still wearing their coveted red berets, told McQuaid and Sigerfoos that "the American airborne boys performed like veterans in their first combat drop."

In that field of blood and pain the feeling was mutual. Private Howard Truitt, of Royal Oak, Michigan, one of Bud Miley's men, declared, "I swear I saw some of the red berets get out of their gliders and take off down the road in their jeeps before the gliders touched the ground."

Many of the men awaiting eventual evacuation were seriously wounded, but McQuaid never heard a whimper or a complaint.

Heavy fighting was taking place only a mile east of the clearing station, and the horizon was filled with smoke and flame. Shells rustled endlessly overhead, and the wounded men flinched as some projectiles exploded less than 100 yards away.

Some of the less seriously wounded raised themselves on their elbows as they spotted an airborne medic walking arduously down the nearby road carrying what at first looked like a child in his arms. But as the trooper got closer the others could see that he was lugging a young German woman. She was strikingly beautiful, but ashen-faced. Her right hand had been shot off that morning after some German snipers hidden in her house opened fire on several 17th Airborne men.

It was nearly dark when generals Ridgway and Miley climbed into a jeep and headed northward to locate the CP of General Eric Bols, leader of the British 6th Airborne. Bols, meanwhile, felt his situation was well enough in hand to relax briefly in his farmhouse headquarters. In the kitchen, festooned with smoked hams, Lieutenant Colonel P. Luard, leader of the 13th Battalion of the 5th Parachute Brigade, cooked scrambled eggs for his divisional commander.

With several troopers "riding shotgun" in a second jeep, Ridgway's and Miley's vehicle cautiously edged through a thick, black forest and reached Bols's CP shortly before 11:00 P.M. It was just past midnight when the two-jeep convoy set out for the return trip to Miley's CP. The jeeps slowed to creep around the burned-out wreckage of a vehicle just as Ridgway caught a glimpse of shadowy figures scurrying about to his front.

"Krauts!" someone called out. Ridgway, Miley, and the others leaped from their vehicle and opened fire. A loud howl rang out, and one of the figures fell to the ground. Ridgway jumped behind a jeep to reload his Springfield, just as an ear-splitting roar erupted. The general felt a burning pain in his shoulder; a grenade had exploded only a few feet from him, but a wheel of the jeep had absorbed much of the blast and shrapnel.

Suddenly, there was eerie silence. Ridgway and Miley could hear men breathing in the blackness on all sides. For long minutes no one moved or uttered a sound. Then Ridgway called out in a stage whisper to Miley, "You okay, Bud? I think I got one of them."

Slipping back into the undamaged jeep, the two generals began edging through the woods once more. Miley saw a movement on the dark path ahead and fired his pistol. There was no response. The 17th Airborne commander, brandishing his Colt .45, crawled out of the jeep and stalked forward until he found one of his men manning a machine gun.

"Damn it!" General Miley exploded spontaneously, "you've got orders to shoot at anything you can't identify! Why didn't you shoot?" Miley—knowing that if the trooper had fired, he would be dead—fought off the temptation to thank his machine-gunner for not following orders in this instance.

At an American casualty station that night, British Captain Boucher-Giles was lying groggily on a cot in a dark tent. That morning he had been shot through the leg while helping unload a glider. He had received several blood transfusions and morphia injections, and was existing in the pitch-black tent in something of a dream world. The clearing station seemed to be right in the center of the battle action, and bullets and fragments from bursting mortar shells occasionally pierced the tents.

Suddenly an especially heavy fusillade of small-arms fire came from Germans in a nearby cluster of trees. Boucher-Giles, raising himself to his elbow, whipped out his pistol and while still on his cot emptied the clip in the direction of the woods. Moments later orderlies rushed to the wounded captain's cot and relieved him of his ammunition; but he refused to give up his pistol.

That night the landscape between the American glider landing zones and the battered city of Wesel to the south was a no-man's land. Guarding a crossroads northeast of Wesel was a group of American glider pilots of the 435th Troop Carrier Group who were organized as a company of infantry. It was nearing midnight when one of the pilots-turned-infantryman, Lieutenant Jacob Zichterman, heard the ominous roar of heavy motors and the clanking of steel treads in the inky blackness to the south. The sounds became louder; Zichterman and the other glider pilots knew that enemy armor was headed their way. The tense pilots were thankful that there were a couple of antiaircraft batteries dug in alongside.

Now heavy bursts of Schmeisser machine-pistol and rifle fire came past the heads of the dug-in pilots. They were under attack by some 150 German grenadiers supported by two or three panzers and self-propelled guns. As coolly as ground-combat veterans, the glider pilots held their fire until the attacking German force was close, then opened up with a barrage of Tommy-gun, machine-gun, and rifle fire. A bazooka lighted the sky with a brilliant orange flame as a rocket swished forward; it exploded near the German tank, causing it to halt, then back up and clank off toward the rear. The enemy foot soldiers were thrown into confusion by the heavy bursts of fire from the pilots, and they pulled back, leaving more than 40 dead comrades.

A short time later two dazed Feldgrau approached the American roadblock to surrender. A glider pilot started to take them prisoner, but a sergeant from the 194th Infantry, who was in command of the position, called out, "We aren't taking prisoners." The pair of Germans were told to start walking down the road and were shot.

When dawn broke it was clear to the Allied high command that Varsity had been a success. All objectives had been secured by the 6th and 17th Airborne Divisions, and the route north of the industrial Ruhr was open for the advance into the heart of the Third Reich. The cost, however, had been heavy. General Bols's Red Devils lost 1,300 men, killed, wounded, or missing (although several hundred of the missing later turned up). General Miley's division had 393 killed, 834 wounded, and 81 missing.

The U.S. IX Troop Carrier Command and the Royal Air Force Tactical command, which had flown the airborne men into battle and then resupplied them, suffered heavily. Total casualties among glider and airplane pilots and crews were 91 killed, 280 wounded, and 414 missing in action.

Late that Palm Sunday morning, the high Allied brass gathered at an ornate château high above the Rhine River, outside Rheinberg. They were all there—American Generals Eisenhower, Bradley, and Simpson; British Prime Minister Churchill and his Field Marshals Brooke and Montgomery. Spirits were soaring. The memory of past disputes was dispelled by the heady intoxicant of victory: three powerful Allied armies were across the Rhine, the last great barrier to the heartland of Germany, and were ready to go forward.

Winston Churchill, who had guided his beleaguered nation through five years of "blood, toil, tears, and sweat," his face flushed with emotion, turned to Supreme Commander Eisenhower and exclaimed, "My dear general, the German is whipped! We've got him! He is through!"[4]

Epilogue

Adolf Hitler's reaction to news that Field Marshal Montgomery's 21st Army Group was over the Rhine in strength was rage. He promptly ordered that poison gas be unleashed against Allied bridgeheads east of the river and that the thousands of American and British airmen who were prisoners of war in the Third Reich be shot. Cooler heads, pointing out that the Western Allies could retaliate on a far more massive scale, prevailed.

Basking in the glow of the Plunder/Varsity triumph, Bernard Montgomery issued a directive for a bold drive toward Berlin. He fired off a signal to his boss and mentor, Field Marshal Brooke: "My goal is to drive hard for the line of the Elbe [River] . . . thence via the autobahn to Berlin, I hope."

Meanwhile, Hodges's and Patton's armies had broken out of bridgeheads at Remagen and at Oppenheim and were racing for a linkup with Simpson's Ninth Army east of the Ruhr industrial basin. To the south, Patch's Seventh Army had bolted across the Rhine at Worms and joined the race toward central Germany.

Joining in the spirit of Allied euphoria, the American Armed

Forces Radio taunted the reeling Wehrmacht by blaring a popular melancholy tune of that period, "I'm Heading for the Last Roundup."

As Field Marshals Brooke and Montgomery and Prime Minister Churchill were eagerly envisioning the capture of Berlin by the Western Allies, Supreme Commander Dwight Eisenhower was burdened with doubts. When the Allies had stormed ashore in Normandy the previous June 6, the ultimate objective had been the nerve center of Nazi Germany—Berlin. But now, in light of dramatic battlefield events of the previous eight weeks, Eisenhower was forced to re-evaluate strategy for the end-the-war smash.

Nine months before, no Allied commander could have foreseen that Hodges's First Army would seize the Remagen Bridge intact, or that Patton would fight through the rugged Palatinate so rapidly and slip across the Rhine, or that the Russians would surge forward 200 miles in only a few weeks and be across the Oder River, only 42 miles from Hitler's bunker in the Reich Chancellery.

General Omar Bradley, perhaps Eisenhower's closest confidant, came to lunch and reinforced the Supreme Commander's growing view that Berlin should not be the ultimate objective of the Western Allies. With Eisenhower's forces still more than 200 miles from Berlin, the Russians would probably get there first, anyhow, Bradley pointed out, adding that President Roosevelt had already conceded the eastern region of Germany to Stalin.

In response to an Eisenhower question, Bradley said, "I think it will cost us one hundred thousand casualties to take Berlin—a pretty stiff price for a prestige objective, especially when we've got to fall back and let the other fellow take over."

Eisenhower and Bradley were convinced that Hitler and other Nazi functionaries would abandon Berlin "in the next two weeks" and flee to the reported Nazi redoubt in the Bavarian Alps, where 250,000 die-hard SS troops would barricade themselves and prolong the war indefinitely. Goebbels's redoubt myth, which had bamboozled SHAEF intelligence for months, was still influencing Allied strategy.

Eisenhower's mind was made up. An attack on Berlin was no longer strategically practical. On March 28, the Supreme Commander issued one of the most fateful directives of the war—and one that created an enormous uproar in high Allied councils. Instead of driving for the Nazi capital, Eisenhower's forces would surround the Ruhr basin, then launch their main attack to the southeast, toward Leipzig and Munich. The Leipzig force was to link up with the Russians as rapidly as possible, slicing Germany in half, while the other troops were to race for the Bavarian and Austrian Alps to wipe out Hitler's redoubt. Bernard Montgomery's Berlin Express would be sidetracked toward the northeast and capture Hamburg and other Baltic ports and prevent German forces from establishing a "northern redoubt" in Norway and Denmark.

That same day, without consultation with his bosses, the Combined Chiefs of Staff, or with the heads of government, General Eisenhower sent a signal to Joseph Stalin, advising the Russian dictator that the Western Allies would attack south of the Nazi capital and leave Berlin to the Red Army.

Stalin was delighted and replied promptly. He assured Eisenhower that the SHAEF plan for concluding the war "entirely coincides with the plan of the Soviet high command."

"Berlin has lost its former strategic importance," Stalin declared. "The Soviet high command therefore plans to allot secondary forces in the direction of Berlin."

At the very hour in which the Russian leader was dictating his reply, he was concentrating five tank armies of 1,500,000 men and 25,000 guns along the Oder River for an assault against the "secondary objective" of Berlin, 40 miles away.

Eisenhower's end-the-war directive and his direct personal contact with a head of state, Stalin, struck the British (and many Americans) with all the delicacy of a blockbuster bomb. Bernard Montgomery was flabbergasted. He was racing for Berlin with the Allied Expeditionary Force's most powerful command and now was told that not only would his thrust be deflected to the northeast but that Bill Simpson's Ninth Army would be taken from him and returned to General Bradley's 12th Army Group for the drive to Leipzig and linkup with the Russians.

The British Chiefs of Staff were shocked to see Eisenhower taking control of the European war as though the Combined Chiefs of Staff did not exist. That night an irate Alan Brooke wrote in his diary:

> To start with, he [Eisenhower] has no business to address Stalin direct. . . . Secondly, he produced a telegram which was unintelligible; and finally, what was implied in it appeared to be entirely adrift and a change from all that had been agreed to.

Without consultation with Churchill, the British Chiefs fired off a telegram to the American Chiefs declaring that Eisenhower had exceeded his authority in contacting Stalin directly and that the strategy leaving Berlin to the Russians was a "grievous political and military blunder." The British Chiefs also stressed that their intelligence apparatus was not concerned in the least with the Hitler redoubt threat and doubted if such an Alpine stronghold existed.

If the British Chiefs of Staff were shocked by the Eisenhower actions, Winston Churchill was stupefied. The supreme commander had made a colossal blunder, he told aides. Churchill now believed that Russia had become "a mortal danger to the free world" and that its sweep to the west should be halted as far east as possible. "Berlin is the prime and true objective of the Anglo-American armies," the prime minister declared.

Eisenhower, aware that he had stirred up a hornets' nest, took pains to account for his decision but stuck to his guns. In a letter to a fretful Montgomery, the supreme commander concluded somewhat angrily: "That place [Berlin] has become, as far as I am concerned, nothing but a geographical location. . . ."

The various thrusts into the Third Reich by several powerful Allied armies moved with great speed. The Russians reached the suburbs of Berlin on April 20, and three days later Omar Bradley's Americans met the Russians at the Elbe River, far west of besieged Berlin. On April 30, with the Red Army spearheads only two blocks away, Adolf Hitler committed suicide in his bunker at the Reich Chancellery. A week later leaders of the decimated Wehrmacht signed unconditional surrender documents. It would be nearly eight weeks before the Russians "allowed" their Western allies to set foot in the conquered capital.

Meanwhile, Allied airborne generals in France and England had been waiting for the signal that never came—the go sign for launching Operation Eclipse, the airborne assault to seize Berlin ahead of the Russians. Long after the war, in 1955, General James Gavin, who would have parachuted into Berlin at the head of his 82nd Airborne Division, was shocked to learn that the Eclipse plan, which had been cloaked in the most intensive secrecy, had been known at the time by Stalin and the Russian high command. Gavin's source was Cornelius Ryan, the noted author of *The Longest Day*, who had interviewed several World War II Russian generals.

One high German officer maintained after the war that the Russians had assembled huge numbers of gliders behind their lines before launching their assaults against Berlin. If this report were true, it could indicate that the Soviets intended to use airborne forces to pounce on Berlin ahead of Eclipse.

During the massive Anglo-American assaults against the Rhine River, one American corps commander, General Manton Eddy, told a confidant, "I've never seen such heat from on high." There was good reason for this "heat": the frightening new threat of the German Messerschmitt-262 jet. If German jet factories were not soon overrun, the entire course of the war could be altered.

Shortly after Germany surrendered, Allied air barons found out that their fears had been justified. General Carl Spaatz and General Hoyt Vandenberg interviewed the corpulent leader of the Luftwaffe, Reich Marshal Hermann Göring, and asked him if the jet warplane really had a chance of winning the war for Germany.

Göring said he was convinced that if Hitler had had only four months' more time, the jet would have won the war. He pointed out that one underground factory, ready for full production when it was seized by American forces, would have produced about 1,200 Messerschmitt jets each month. "With five or six thousand jets, the outcome would have been quite different," Göring declared. The Luftwaffe leader added that there was a surplus of jet pilots, and that underground factories had been set up for oil.

In the realm of what might have been, German generals after the war maintained that had General Simpson been allowed by Field

287

Marshal Montgomery to take the Rhine on the run when Simpson first reached the river in early March, there would have been only scattered, exhausted men, desperately short of arms and ammunition, to oppose the Ninth Army crossing.

The dash over the Remagen bridge by an American infantry platoon to seize the span and shorten the war was hailed as among the most courageous—and significant—actions in military history. Lieutenant Karl Timmermen and a few of his men received the Distinguished Service Cross. But their deeds were soon forgotten by much of the American public. After the war, Lieutenant Timmerman returned to his home town of West Point, Nebraska, a lonely figure trudging along with a barracks bag over one shoulder. His reception committee was a mongrel dog that snarled and snapped at his heels. Timmerman rejoined the army in 1948, saw action in the Korean War, and died at an early age of cancer, in 1951.

Bitter wartime personality and professional clashes among American and British leaders spilled over into the postwar era. President Roosevelt's secretary of the treasury, Henry Morgenthau, who created the plan to turn Germany into farmland and execute thousands of German officers and officials, wrote a book in which he described a lunch he had with Eisenhower in Portsmouth, England, on August 7, 1944, as the source of his plan for the brutal treatment of a subjugated Germany. Eisenhower, purple with rage, vehemently denied this and never forgave Morgenthau. Late in 1944, the supreme commander had called the treasury secretary's proposals "criminal."

Bernard Montgomery in his memoirs declared with characteristic candor that Eisenhower as a field commander was "very bad, very bad." Stung by the caustic remark, Eisenhower lashed back. Montgomery was "a psychopath," he charged. The former supreme commander ceased corresponding with Montgomery, claiming he "had no interest in keeping up contact with a man who just can't tell the truth."

Omar Bradley categorized Montgomery as "detestable" and said that he never spoke to the field marshal either. Alan Brooke published his memoirs a few years later. So pointed were many of his remarks concerning Winston Churchill that the wartime prime minister refused ever to see or speak to Brooke again.

Churchill, who had guided Great Britain through five turbulent and perilous years of World War II, who had been a beacon of inspiration to the free world during the period when the island nation stood alone against Hitler's powerful legions and Luftwaffe, received curious recognition from his people. A few months after the shooting stopped in Europe, they voted him out of office.

Notes and Sources

Introduction
 1. Frederick Ayers, *Patton*.

Chapter 1. A Dispute Among Generals
 1. John Toland, *The Last 100 Days*.
 2. Ibid.
 3. Anthony Cave Brown, *Bodyguard of Lies*.
 4. John Toland, *The Last 100 Days*.
 5. Ibid.
 6. Omar N. Bradley, *A Soldier's Story*.
 7. Ladislas Farago, *Patton—Ordeal and Triumph*.
 8. Omar N. Bradley, *A Soldier's Story*.
 9. Ibid.
10. John Toland, *The Last 100 Days*.

Chapter 2. Hitler's Looming Wonder Weapons
 1. Charles H. McDonald, *The Last Offensive*.
 2. John Toland, *The Last 100 Days*.

Chapter 3. An Angry Patton Threatens to Resign

1. Ladislas Farago, *Patton*.
2. Two months later, with his entire Army Group B surrounded in Germany's Ruhr, Feldmarschall Model killed himself with his Luger.
3. Albert Speer, *Inside the Third Reich*.

Chapter 4. "Sleeper Patrols" Among the Enemy

1. Quoted in *Time* magazine.
2. Leo A. Hough and Howard J. Doyle, *Timberwolf Tracks*.
3. Ibid.

Chapter 5. Assault Over the Roer.

1. Theodore Draper, *84th Infantry Division*.
2. J. Lawton Collins, *Lightning Joe*.
3. Author's diary.
4. Most American soldiers who fought throughout the war in Europe agreed that Düren was the most nearly pulverized city they had seen.

Chapter 6. Hitler Rejects a Pullback Over the Rhine

1. Author's diary.
2. Leo A. Hough and Howard J. Doyle, *Timberwolf Tracks*.
3. In the late 1950s, as governor of Arkansas, Orval Faubus clashed violently with his World War II supreme commander when President Eisenhower sought to enforce school integration at Little Rock after a landmark Supreme Court decision abolishing racial discrimination in public learning institutions.
4. Charles H. McDonald, *The Last Offensive*.
5. Leo A. Hough and Howard J. Doyle, *Timberwolf Tracks*.
6. This episode was related to the author by Haynes W. Dugan of Shreveport, Louisiana, wartime assistant intelligence officer of the 3rd Armored Division.

Chapter 7. Vaulting the Erft Canal

1. Walther Warlimont, *Im Hauptquartier der deutschen Wehrmacht*.
2. Episode related to author by Haynes W. Dugan. Added details in Frank Woolner's *Spearhead in the West*.

Chapter 8. Duel at Adolf Hitler Bridge

1. *Time* magazine.

2. After the war, most German commanders agreed: a surprise crossing by Simpson's Ninth Army at Düsseldorf would have met little opposition.
3. Theodore Draper, *84th Infantry Division*.
4. Ladislas Farago, *Patton*.
5. Interview with Haynes W. Dugan.
6. J. Lawton Collins, *Lightning Joe*.

Chapter 9. Bloody Road to the Wesel Bridges
1. Charles H. McDonald, *The Last Offensive*.
2. John Toland, *The Last 100 Days*.
3. Leo A. Hough and Howard J. Doyle, *Timberwolf Tracks*.
4. J. Lawton Collins, *Lightning Joe*.

Chapter 10. "Mighty Cologne Has Fallen"
1. Frank Woolner, *Spearhead in the West*.
2. Author's diary.
3. Leo A. Hough and Howard J. Doyle, *Timberwolf Tracks*.
4. Most American soldiers went through the war holding the opinion that the term for these lethal mines was "shoe" because they often blew off a man's foot. Actually, Schu mines were named after their German inventor, Dr. Schu.
5. To the present day, veterans of the 3rd Armored Division and 104th Infantry Division claim that their respective outfits were the first to enter Cologne.
6. The fate of General Köchling was never officially determined.
7. Interview with *Chicago Tribune* war correspondent John H. "Beaver" Thompson. He was the first American reporter to jump with a paratrooper unit on a combat operation, in North Africa during the 1942 Allied invasion there.
8. Leo A. Hough and Howard J. Doyle, *Timberwolf Tracks*.
9. J. Lawton Collins, *Lightning Joe*.
10. Leo A. Hough and Howard J. Doyle, *Timberwolf Tracks*.

Chapter 11. "Your Name Will Go Down in Glory!"
1. Actions and events in the dramatic Remagen Bridge episode were pieced together by the author from numerous sources, including official U.S. Army accounts and battle after-action reports. But by far the principal source for recounting the events and dialogue was *The Bridge at Remagen* by Kenneth Hechler (1957). The author is indebted to Ballantine Books

for permission to draw extensively on the dialogue. Mr. Hechler, who was an Army historian during the war, later became a member of the U.S. House of Representatives.

Chapter 12. "Suicide Mission" at Ludendorff Bridge

1. Omar N. Bradley, *A Soldier's Story*.
2. Ibid.
3. Ibid.
4. Ibid.
5. Dwight D. Eisenhower, *Crusade in Europe*.

Chapter 13. Hitler Thirsts for Scapegoats

1. Joseph B. Mittelman, *Eight Stars to Victory*.
2. Ibid.
3. Albrecht Kesselring, *A Soldier's Record*.
4. Omar N. Bradley, *A Soldier's Story*.

Chapter 14. "Annihilation of Two Armies Is Imminent"

1. Charles H. McDonald, *The Last Offensive*.
2. Ladislas Farago, *Patton*.
3. George Dyer, *XII Corps, Spearhead of Patton's Army*, privately printed.
4. Ladislas Farago, *Patton*.
5. Charles H. McDonald, *The Last Offensive*.
6. George Dyer, *XII Corps, Spearhead of Patton's Army*.
7. Ibid.

Chapter 15. Patton: "Holding the Hun by the Nose"

1. By war's end, General Robert T. Frederick had been wounded nine times.
2. J. Lawton Collins, *Lightning Joe*.
3. Supreme Commander Dwight Eisenhower was greatly impressed by the bridge-building feat. He told Chief of Staff George C. Marshall the story in a letter, but raised the ante from beer to champagne. This caused General Collins to tell his aides, "I couldn't afford champagne for an entire battalion."
4. *Time* magazine.
5. Major General Millikin was not reduced in rank, but given command of the 13th Armored Division, which he led until the end of the war.

Chapter 16. "For God's Sake, Tell Them I'm Across!"

1. Charles H. McDonald, *The Last Offensive*.
2. Ladislas Farago, *Patton*.
3. Ibid.
4. Ibid.
5. George Dyer, *XII Corps, Spearhead of Patton's Army*.
6. Samuel Eliot Morison, *The Invasion of France and Germany*.
7. George Dyer, *XII Corps, Spearhead of Patton's Army*.
8. Omar N. Bradley, *A Soldier's Story*.
9. Ibid.
10. John Toland, *The Last 100 Days*.

Chapter 17. A Daring Plan to Seize Berlin

1. Much of the detail as well as the general outline of this plan to seiz[e] Berlin ahead of the Russians was furnished the author by General Jame[s] Gavin in 1984.
2. Arie Bestebreurtje became an American citizen after the war. He died b[y] drowning in 1983 after many years as a clergyman. General James Gavi[n] told the author that Bestebreurtje was "one of the bravest men I kne[w] during the war."
3. The Russians kept their Allies—the Americans, British, and French— out of Berlin for 53 days after V-E Day. During that time the Russian[s] set up much of the political machinery that has confronted the Wester[n] Allies since that period.
4. Colonel Barney Oldfield of Los Angeles, who would have led the corre[-] spondents in the proposed parachute assault on Berlin, provided the au[-] thor with many details on the operation.
5. Asked by the author in 1984 if the parachute assault on Berlin could hav[e] been a "suicide mission," General James Gavin replied noncommittally[:] "It would have been an interesting operation."
6. So tight was the secrecy that many junior officers who trained for th[e] Berlin jump were unaware for many years after the war that the Naz[i] capital was the target, they have told the author.
7. Albert Speer, *Inside the Third Reich*.
8. Ibid.

Chapter 18. "Two If by Sea"

1. Colonel Edson Raff's unit at that time was officially the 2nd Battalion

503rd Parachute Infantry Regiment. Its designation was later changed to 509th Parachute Infantry Battalion.

2. General William Miley told the author in 1984 that he was not in favor of the daylight airborne assault, but that there were so many complex factors involved that he finally agreed to it.

3. Charles B. McDonald, *The Last Offensive*.

4. Ibid.

Chapter 19. Crossing the Moat by Buffalo

1. John Toland, *The Last 100 Days*.

2. Field Marshal Alan Brooke, *The Alan Brooke Diaries*.

3. A Buffalo was an amphibious craft which could be driven on land.

4. 30th Infantry Division history.

5. Ibid.

6. Dwight D. Eisenhower, *Crusade in Europe*.

7. George Dyer, *XII Corps: Spearhead of Patton's Army*.

8. Ibid.

9. Ladislas Farago, *Patton*.

10. Omar N. Bradley, *A Soldier's Story*.

Chapter 20. The Paratroopers Strike

1. A "stick" was an arbitrary number of paratroopers, normally 15 to 18, jumping out of an aircraft.

Chapter 21. Dropping Onto a Hornets' Nest

1. General James M. Gavin to the author.

2. General Miley told the author in 1984 that, despite his difficulties on landing, within an hour and a half he felt that he had established control of his division, a very short time for a major airborne operation.

Chapter 22. The C-46 Deathtraps

1. To the astonishment of his colleagues, Robert Capa survived the war. Some years later he was killed while covering front-line action in a war in Indochina.

2. The wounded paratrooper survived and in 1984 was still going strong.

3. Captain Charles Jones later rose to major general in the Ohio National Guard.

Chapter 23. Glider Fighters Pounce

1. Frank Guild, Jr., *Action of the Tiger*.
2. Ibid.
3. Years after the war, Vermillion was on the night desk of United Press in New York City when the telephone rang. A British voice said, "I say, old boy, I'm awfully sorry about the typewriter, and I want to tell you what happened. This is Jeff Bocca." He told of his ordeal, then added, "I did try, Bob, but really, your typewriter was quite full of holes."
4. British Air Ministry, *By Air to Battle*.

Chapter 24. "General, the German Is Whipped!"

1. Before entering the army, Major Barney Oldfield was a Hollywood press agent. Among his clients were a rising young actor named Ronald Reagan and a child prodigy named Elizabeth Taylor.
2. As reported in *Time* magazine.
3. Dwight D. Eisenhower, *Crusade in Europe*.
4. Ibid.

Bibliography

Books

Ambrose, Stephen E. *The Supreme Commander*. New York: Doubleday, 1970.

Ayers, Frederick. *Patton*. Boston: Little, Brown, 1972.

Bauer, Eddy. *Encyclopedia of World War II*. New York: Marshall Cavendish Corp., 1970.

Bekker, Cajus. *The Luftwaffe War Diaries*. New York: Doubleday, 1969.

Bennett, Ralph. *Ultra in the West*. New York: Charles Scribner's Sons, 1979.

Blumenson, Martin. *The Patton Papers*. Boston: Houghton Mifflin, 1974.

Blummentritt, General Guenther. *Von Rundstedt*. London: Oldham's, 1952.

Bradley, General Omar N. *A Soldier's Story*. New York: Henry Holt & Co., 1951.

Brooke, Alan, the Viscount. *Diaries*. London: Collins, 1957.

Brown, Anthony Cave. *Bodyguard of Lies*. New York: Harper & Row, 1975.

Buckmaster, Maurice. *They Fought Alone*. New York: W. W. Norton, 1958.

Butcher, Harry C. *My Three Years With Eisenhower*. New York: Simon & Schuster, 1946.

Chatterton, George. *Wings of Pegasus*. London: McDonald, 1962.

Churchill, Winston. *Memoirs*. Boston: Houghton Mifflin, 1948.

Collins, General J. Lawton. *Lightning Joe*. Baton Rouge: Louisiana State University, 1979.

Eisenhower, General Dwight D. *Crusade in Europe*. New York: Doubleday, 1948.

Farago, Ladislas. *Patton—Ordeal and Triumph*. New York: Obolensky, 1964.

Faubus, Orval E. *In This Faraway Land*. Little Rock: River Road Press, 1971.

Garlinski, Jozef. *Hitler's Last Weapons*. New York: Times Books, 1978.

Gavin, General James M. *On to Berlin*. New York: Viking, 1978.

Hechler, Kenneth. *The Bridge at Remagen*. New York: Ballantine Books, 1957.

Horrocks, General Brian. *A Full Life*. London: Leo Cooper, 1974.

Irving, David. *The War Between the Generals*. New York: Congdon & Lattes, 1981.

Keitel, Field Marshal Wilhelm. *The Memoirs of Field Marshal Keitel*. New York: Stein & Day, 1965.

Kesselring, Field Marshal Albrecht. *A Soldier's Record*. 1954.

Killen, John. *A History of the Luftwaffe*. New York: Doubleday, 1968.

Lewin, Ronald. *Ultra Goes to War*. New York, McGraw-Hill, 1978.

Majdalany, Frederick. *The Fall of Fortress Europe*. New York: Doubleday, 1968.

McDonald, Charles B. *The Last Offensive*. Washington: Chief of Military History, 1973.

Montgomery, Field Marshal Bernard. *Memoirs*. London: Collins, 1958.

Morison, Samuel Eliot. *The Invasion of France and Germany*. Boston: Little, Brown, 1957.

Mrazek, James E. *The Glider War*. New York: St. Martin's Press, 1975.

Oldfield, Colonel Barney. *Not a Shot in Anger*. New York: Duell, Sloan & Pearce, 1956.

Patton, General George S. *War As I Knew It*. Boston: Houghton Mifflin, 1947.

Ridgway, General Matthew B. *Soldier*. New York: Harper, 1956.

Speer, Albert, *Inside the Third Reich*. New York: Macmillan, 1962.

Summersby, Kay. *Eisenhower Was My Boss*. New York: Prentice-Hall, 1948.

Taylor, General Maxwell D. *Swords and Plowshares*. New York: W. W. Norton, 1972.

Thompson, R. W. *Men Under Fire*. London: Macdonald, 1945.

Toland, John. *The Last 100 Days*. New York: Random House, 1966.

Warlimont, General Walther. *Im Hauptquartier der deutschen Wehrmacht*. Frankfurt: Bernard und Graefe Verlag Wehrwesen, 1962.

Westphal, General Siegfried. *The German Army in the West*. London: Cassell, 1951.

Winterbotham, F. W. *The Ultra Secret*. New York, Harper & Row, 1974.

UNIT HISTORIES

British Air Ministry. *By Air to Battle*. London: His Majesty's Stationery Office, 1946.

Dawson, W. Forest. *Saga of the All American*. Privately printed, 1946.

Draper, Theodore. *84th Infantry Division*. New York: Viking Press, 1946.

Dyer, George. *XII Corps: Spearhead of Patton's Army*. Washington: privately printed, 1946.

Guild, Frank, Jr. *Action of the Tiger*. Nashville: Battery Press, 1980.

Hough, Leo A., and Doyle, Howard J. *Timberwolf Tracks*. Washington: Infantry Press, 1946.

Mason, S. B. *Danger Forward*. Atlanta: Albert Love Enterprises, 1947.

Mittelman, Joseph B. *Eight Stars to Victory*. Washington: Infantry Press, 1948.

Pay, Don R. *Thunder from Heaven*. Birmingham, Mich.: *Boots, The Airborne Quarterly*, 1947.

VII Corps' History of Operations in Europe. Privately printed, 1945.

78th Infantry Division Association, *Lightning*. Washington: Infantry Press, 1947.

Warren, John C. *Airborne Operations in World War II*. Maxwell Air Force Base: USAF Historical Division, 1956.

Woolner, Frank. *Spearhead in the West*. Privately printed, 1946.

PAMPHLETS

Davis, Brian L. *German Parachute Forces*. New York: Arco, 1974.

Davis, Howard P. *British Parachute Forces*. New York: Arco, 1974.

PERIODICALS

Collier's, Life, Time, Saturday Evening Post magazines.

Chicago Tribune, New York Times, St. Louis Globe-Democrat newspapers.

MISCELLANEOUS

Combat interviews, U.S. Military History Institute, Carlisle Barracks, Pennsylvania:

Generalleutnant Fritz Bayerlein (German).
Major General John Leonard, 9th Armored Division, Remagen bridgehead.
Generalleutnant Eugen Meindl (German).
Major General John Millikin, III Corps, Remagen bridgehead.
Colonel James H. Phillips, Chief of Staff, III Corps.
Lieutenant Colonel Clayton A. Rust, 276th Engineer Combat Battalion, Remagen Bridge.

Index

Admiral Scheer Bridge, 102–104
Adolf Hitler Bridge, 85–89
"aggressive defense," 18, 32–33, 70, 143
Ahrnsbrak, Francis, 47
Ahroon, Lester A., 77
Alderdyce, C. W., 261
Allen, Terry de la Mesa, 36, 54, 67, 78,
  111, 161
Allied commanders, disputes among,
  3–12, 17–18, 32, 83–84, 90,
  141–145, 156, 199–201, 208–210,
  281, 285–286, 288–289
Alpine Redoubt, 12–13, 14, 284, 286
Anderson, Andy, 269
Anderson, John B., 98
Anderson, K. A. N., 210
Andrus, Clift, 150
Anglo-American Combined Chiefs of
  Staff, 6–10, 275, 285, 286
Army Air Corps, U.S., 160
Army Group B, German, 33, 119, 165
Army Group G, German, 166

Baade, Leonard, 63, 98, 99
Baal, 56–58
Bad Godesberg, 150–151

Baerl Bridge, 102–103
Baldwin, Robert M., 236–237
Banbury, Robert G., 115
Barkley, Harold E., 232, 235
Barr, Andrew, 92
Barton, Robert, 37
Baruch, Bernard, 5–6
Bates, James, 113
Battle of the Bulge, 17, 18, 19, 142, 153,
  171, 225, 229, 230, 247, 257
Baur, Friedrich, 216
Below, Nickolaus von, 218
Berkowitz, Morris, 224
Berlin:
  as military objective, 192–197, 239,
    283–287
  as surrounded by Russian army, 3–4,
    195, 197, 284
"Berlin Bitch," 211–212, 261
Bestebreurtje, Arie, 192–194
Binek, Eugene J., 114–115
Bismarck, Otto von, 55
Blaskowitz, Johannes, 202
Blunck, Stanley R., 68
Bocca, Geoffrey, 228, 268
Boles, Thomas, 96–97

Bolling, Alexander R., 45, 51, 72, 86–87, 102

Bols, Eric, 205, 207, 208, 209, 243–244, 278

Bonn Bridge, 151–152

Booth, Kenneth L., 254–255

Bormann, Martin, 5

Botsch, Walther, 119–120, 125

Boucher-Giles, Captain, 270, 279

Bourne, G. K., 267–268

Bower, Hayden, 54, 55

Boyle, Harry, 253–254, 256–257

Bradley, Omar, 29, 32, 83, 286, 273, 281
   Eisenhower's relations with, 11–12, 14, 17–18, 284
   Montgomery's relations with, 17–18, 90, 153, 225–226, 288
   Rhine strategy of, 7, 8, 10, 11, 141–145, 156, 165, 178–179, 189–190, 275

Brandenberger, Eric, 33, 69

Branigan, Edward S., Jr., 234, 241–242

Bratge, Willi, 118–121, 125–126, 135, 139, 159

Braun, Wernher von, 24

Brereton, Lewis H., 206

Brightman, Sam, 273

Brooke, Alan, 7, 8, 9, 143, 197, 204, 219, 234, 281, 283, 284, 286, 288–289

Brown, Nigel, 270–271

Bruske, Siegfried von, 86

Buck, William C., 111

Bull, Harold "Pink," 140–145, 156

Burgdorf, Fritz, 218, 219

Burleson, John, 107

Burr, Herbert J., 182

C-46s, 247–258

C-47s, 206, 229, 265–266

Campbell, Doon, 227

Capa, Robert, 248, 249

"Captain Harry," 192–194

Carver, Leroy, 51

Church, John H., 73

Churchill, Winston S., 5, 6, 8–9, 23, 143, 197, 203–204, 219, 225, 234, 281, 284, 286, 288–289

Clawson, Allen, 49

Coates, Charles E., 93

Codman, Charles, 225

Coleman, Randolph, 117

Collins, J. Lawton "Lightning Joe," 22, 45, 53–54, 58, 75–79, 90, 92, 105, 106, 114, 116, 160–161, 173–174

Cologne:
   Allied ground assault on, 68, 97, 105, 106–117
   bombing of, 76–77, 106

Colville, Jock, 204

Coningham, Arthur, 206

Conn, Paul, Jr., 188–189

Cook, Robert J., 112, 113

correspondents, war, 12–13, 195, 199–201, 272–274

Cota, Norman D. "Dutch," 25

Cothran, Ben, 122, 149

Coutts, James W. "Lou," 247, 248–250, 255, 257, 271

Craig, Lou, 147, 150, 154, 158

Crerar, Henry D. G., 20, 22, 66, 98

Crockett, Kyser, 73

Crouch, Cleo I., 231–232

Crouch, Joel L., 234

Cummings, Joseph M., Jr., 78

Dabrowski, Chester, 221–222

Dager, Holmes, 161–162, 182

Dahlquist, John E., 171

Das, Stanley, 222

Deer, Victor, 262–263

Deevers, Murray, 133, 136, 139, 146, 147

Dempsey, Miles, 213–214

De Rango, Fred, 128, 130

Deutsch, Friedrich, 221

DeLisio, Joseph, 126–129, 137–138

Dobie, Norman D., 73

Dorland, Eugene, 134

Dornberger, Walther, 24

Dove, George, 104

Doyle, John K., 248

Drabnik, Alex, 137, 146

Dubskey, Joe, 264

Dugan, Haynes W., 91–92, 114

Dulles, Allen, 190–191

Düren, 54–56

Dwelly, Robert C., 43–44

Early, Robert, 113–114

Earnest, Hubert L., 166

Eckhoff, William, 50–51

Eclipse, Operation, 192–197, 239, 287

Eddy, Manton S. "Matt," 167, 181–182, 185, 186, 287

Eighth Air Force, U.S., 224

8th Armored Division, U.S., 98, 99–102
8th Infantry Division, U.S., 45, 68, 77, 78, 91, 105
8th Parachute Division, German, 73
XVIII Airborne Corps, U.S., 199, 214
82nd (All American) Airborne Division, U.S., 193, 194, 196
83rd Armored Reconnaisance Battalion, U.S., 92–93
83rd Infantry Division, U.S., 81, 84
84th (Railsplitters) Infantry Division, U.S., 45, 51–52, 56–58, 72–75, 85, 86, 102–103
LXXXVI Corps, German, 243
86th Engineer Pontoon Battalion, U.S., 156
87th Mortar Battalion, U.S., 108
Eisenhower, Dwight D., 82–83, 172–173, 178
  Bradley's relations with, 11–12, 14, 17–18, 284
  Rhine strategy of, 6–8, 32, 141–145, 155, 156, 163, 179, 190, 234, 275
  as Supreme Commander, 11–12, 13, 44, 96, 116, 197, 222–223, 281, 284–286, 288
11th Armored Division, U.S., 182
Elmer, Warren P., 255
Engeman, John, 123, 124, 126, 128, 130, 133–134, 140, 146–147
Enigma coding machine, 28–29
Erft Canal complex, 68–69, 71–80, 90, 94
Erft Dam, 19–20
Erpeler Ley, 119, 125, 126, 138, 146, 175
Erwin, Delbert, 37
Evers, Jacob L., 7

Faubus, Orval E., 64–65
Faust, Sergeant, 136
Felber, Hans, 33, 69, 70, 165–166, 181
Fifteenth Army, German, 121
5th Armored Division, U.S., 66
5th Infantry Division, U.S., 161, 166–167, 185, 186, 189, 223
XV Corps, U.S., 171
51st Highland Division, British, 219–220, 245–246
LIII Corps, German, 120
Filburn, Daniel, 98
First Army, Canadian, 7, 20, 66
First Army, French, 7, 170
First Army, German, 181, 183

First Army, U.S., 7, 10, 11, 19, 22, 31, 90, 93, 143, 149–150, 159, 165, 179, 274, 283, 284
1st Commando Brigade, British, 220
1st (Big Red One) Infantry Division, U.S., 150, 151, 152
507th (Raff's Ruffians) Parachute Infantry Regiment, U.S., 231, 234, 242, 243
509th Parachute Infantry Battalion, U.S., 114
513th Parachute Infantry Regiment, U.S., 247, 250, 252, 255–256, 257, 271
555th Antiaircraft Artillery Battalion, U.S., 60
557th Antiaircraft Artillery Battalion, U.S., 61
Flying Special Tribunal West, 155, 158, 159–160
Foertsch, Hermann, 170, 181
Foster, Hubert, 107
4th Armored Division, U.S., 134, 161, 162, 165, 166, 167, 183–184, 190
45th (Thunderbird) Infantry Division, U.S., 171, 183
47th (Raiders) Infantry Regiment, U.S., 147–148
407th Infantry Regiment, U.S., 52–53
413th Infantry Regiment, U.S., 37–38
414th Infantry Regiment, U.S., 36, 38, 78, 104, 111
415th Infantry Regiment, U.S., 108, 111
435th Troop Carrier Group, U.S., 279–280
437th Troop Carrier Group, U.S., 228, 229
464th (Branigan's Bastards) Parachute Field Artillery Battalion, U.S., 234, 241–242
466th Parachute Field Artillery Battalion, U.S., 247, 254
Frederick, Robert T., 171, 183
Friesenhahn, Karl, 120–121, 129, 135–136, 139, 160, 176

Gaffey, Hugh, 161, 163–165, 167, 172
Gale, Richard, 212
Gallagher, Wes, 200
Garcia, Manuel, 96
Gavin, James M., 144, 194, 196, 239–240, 287
Gay, Hobart "Hap," 184

Gerhardt, Charles H., 45, 53, 81–82
Gerow, Leonard, 185
Gersdorff, Christoff Georg Count von, 70
Gibble, William T., 137, 149
Gildersleeve, Francis, 266
glider assaults, 206, 259–271
Goebbels, Josef, 4, 13, 211, 218, 219, 284
Golzheim, 67–68
Goodman, Francis, 176
Göring, Hermann, 14, 16, 153, 287
Gott, Rod, 186, 189
Grasshopper, Operation, 186, 189
Green, Robert D., 95–96
Gregoire, Alphie, 108
Grenade, Operation, 20, 22, 27–28, 32, 38, 42, 44–58, 67
Grigg, James, 180
Grimball, John, 124, 128, 129, 133–134
Groves, James "Red," 55
Growdon, John "Pinky," 123
Guderian, Heinz, 14, 16
Gunn, Frank L., 150
Gurney, Edward, 100–102

Hagerman, Bart, 257–258
Haislip, Wade H., 171
Hale, George H., 51–52
Hart, Charles E., 22
Hartments, John H., 98
Hausser, Paul, 165, 181
Haynes, Tommy, 186, 189
Henderson, J. R., 246
Herrera, Silvestre, 171–172
Hiatt, Charles F., 75
Hiese, Thornton E., 265
Hill, James, 208, 245
Himmler, Heinrich, 13, 65
Hinds, Sydney, 87–88, 89
Hitler, Adolf, 14–16, 29, 33–34, 284, 286
    health of, 14–15, 23
    loyalty to, 82, 112, 191, 211
    Rhine strategy of, 5, 12–13, 19, 23–25, 34, 60, 65, 67, 70, 71–72, 105, 135, 141, 152–155, 168, 175, 180, 184, 204, 217–219, 283
Hitzfeld, Otto, 121, 125
Hobbs, Leland S., 45, 53
Hodges, Courtney H., 10, 11, 18–19, 20, 22, 90, 121, 141–142, 154, 156, 158–159, 177–178, 190, 191, 225, 275
Hogan, Samuel, 69, 79
Hoge, William M., 122, 123, 125, 128,

130, 133, 135, 139–140, 141, 148–149
Hohenzollern Bridge, 97, 105, 111
Hooker, Jerry, 46–47, 49
Hoover, Fred, 110–111
Horrocks, Brian G., 20–22, 26–27, 31, 66, 246
Hottelet, Richard C., 238–239, 273
Howdieshell, Harold L., 75
Hübner, Rudolf, 155, 158, 159
Huebner, Clarence R., 19, 22, 179

intelligence, military, 12–13, 25, 28–29, 45, 284
Irwin, S. Leroy "Red," 166–167, 185, 187, 190

Jacobs, Irven, 187–188
Jenkins, J. H., 268–269
Jensen, Marvin, 137
Jensen, Roland, 167
Jodl, Alfred, 15–16, 152, 153
Johnson, Chuck, 229
Jones, Charles H., 254

Keating, Frank A., 45, 52
Keitel, Wilhelm, 15
Kelleher, Gerald C., 108
Kelly, David B., 100
Kesselring, Albrecht, 152, 154–155, 157–158, 160, 165, 166, 180, 181, 184, 190, 191, 204, 217, 226
Kimball, Edward A., 99, 101–102
Kissinger, Edmund L., 242
Koblenz, 166, 168–169
Köchling, Friedrich, 97, 111
Koniev, Ivan, 197
Kortzfleisch, Joachim von, 140–141
Kraft, August, 159–160
Krefeld, 85–86

Lanier, Henry L., 64–65
Larsen, Harold, 125
Lauer, Walter E., 78
LeFevre, Brack, 262–263
Leide, William, 186–187, 189
Leigon, Walter, 61, 62, 95
Leonard, John W., 121, 122, 124–125, 134, 139–140, 142
Ley, Robert, 4
Lieberman, Morton R., 49–50
Longest Day, The (Ryan), 287

Loringhoven, Bernd Freytag von, 14
Lovelady, William B., 79, 107
Ludendorff Bridge, 118–155, 157,
    159–160, 173, 175, 176–178, 184,
    288
Luftwaffe, 24, 44, 59–61, 149, 214

McGloin, Joe, 266
McIlwain, William M., 39–42
McLain, Raymond S., 89
Macon, Robert L., 81
MacParland, Eugene, 117
McQuaid, B. J., 277, 278
Magill, John F., Jr., 210–211, 228,
    247–248, 255
Maness, Lewis, 148
Manteuffel, Hasso von, 19
Marcyyk, John, 264
Marino, Anthony, 224
Market-Garden, Operation, 193, 205
Marshall, George, 8–10
Martin, Mike, 117
Massie, Art, 137
Maximilian Bridge, 184
May, Andrew J., 274
Meindl, Eugen, 214–216
Messerschmitt jet fighter planes, 24, 44,
    59–61, 83, 287
Milas, John, 114, 115
Miley, William M. "Bud," 144, 205, 207,
    208–209, 229–230, 232, 240, 257,
    275, 278–279
Military Police, U.S., 153–154
Miller, A. C., 250
Miller, C. Windsor, 147
Miller, Graham, 55–56
Millikin, John, 122, 159, 177–178
Model, Walther, 33–34, 65–66, 69–70,
    95, 118, 140, 141
Moir, William W., 250–252
Montgomery, Bernard, 20, 32, 197, 288
    Bradley's relations with, 17–18, 90,
        153, 225–226, 288
    Operation Plunder directed by, 179,
        180, 185, 190, 191, 198–204, 208,
        213, 216, 219, 281
    proposed Berlin drive of, 283, 284, 285,
        286
    Rhine strategy of, 7–10, 11, 12, 20, 29,
        32–33, 66, 83–84, 88–90, 97, 143,
        144, 145, 156, 196, 287–288
Moore, Bryant E., 78
Morell, Theodor, 15, 23

Morgan, Frederick, 144
Morgenthau, Henry, 288
Morris, Lew, 263
Mott, Hugh, 134, 137
Murphy, Frederick C., 182–183
Mussolini, Benito, 153

Nagel, Gordon L., 235–236
Napoleon I, Emperor of France, 3, 185
Naval Unit 2, U.S., 186, 189
Neilsen, Robert, 46
Nelson, William, 56–57
Neptune, Operation, 7, 205
Nicklin, J. S., 244
Ninth Air Force, U.S., 232
9th Armored Division, U.S., 121,
    122–123, 175
Ninth Army, U.S., 7, 11, 12, 17–18, 19,
    22, 28, 31, 58, 81, 89, 121, 180–181,
    200–202, 225, 272–273, 283, 285,
    287–288
9th Infantry Division, U.S., 147–148,
    150, 151, 158
IX Troop Carrier Command, U.S., 206,
    231, 280
XIX Corps, U.S., 89
90th Infantry Division, U.S., 166, 167
92nd Armored Field Artillery Battalion,
    U.S., 87
95th Infantry Division, U.S., 11
99th Infantry Division, U.S., 78, 90
991st Armored Field Artillery Battalion,
    U.S., 69

Oberkassel, 84–85
Odorizzi, "Odo," 251–252
Oldfield, Barney, 195, 199, 200, 272–274
Oliver, James, 246
Oliver, Lunceford E., 66
Olson, Sidney, 66–67
Ondrick, John G., 29–30
O'Neal, Joel, 238, 239
101st (Screaming Eagles) Airborne
    Division, U.S., 196
102nd Infantry Division, U.S., 45, 52
104th (Timberwolf) Infantry Division,
    U.S., 36–37, 38–42, 45, 46, 49–51,
    61–63, 67–68, 78, 91, 95–97,
    104–105, 106, 108, 111, 116–117,
    161
139th Airborne Engineer Battalion, U.S.,
    264
142nd Infantry Regiment, U.S., 172

194th Glider Infantry Regiment, U.S., 229, 232, 263, 267
1058th Bridge Construction and Repair Group, U.S., 177
Oppenheim, Waldemar von, 94–95
Oriola, Ralf Count von, 70
OSS (Office of Strategic Services), 190–191, 192–193

Parker, Edwin P., Jr., 19, 148
Parker, Harry F., 98–99
Parker, Warme, 221
Patch, Alexander M. "Sandy," 163, 166, 170–171, 181
Patton, George Smith, Jr., 10–11
  reputation of, 9, 67, 90, 201, 274–275
  Rhine strategy of, 18, 32–33, 69, 70, 120, 125, 156, 161, 163–168, 171–173, 179, 181–182, 184, 185, 189–190, 225–226, 283
Pierce, James R., 260, 263, 267
Plunder, Operation, 142, 144, 156, 179, 198–204, 207, 217
  as directed by Montgomery, 179, 180, 185, 190, 191, 198–204, 208, 213, 216, 219, 281
  see also Varsity, Operation
Poett, James, 245
Porter, William, 276
Posey, Andrew, 37–38
Prince, William R., 123

Quesada, Elwood R. "Pete," 224

Radio Berlin, 34, 149, 175, 176, 207, 211–212
Radio Luxembourg, 149
Raff, Edson D., 209–210, 231–235, 241, 242–243, 247
Randle, William, 188
Rankin, John E., 274
Rees, 219–220, 245–246
Reichhelm, Günther, 140
Remagen, see Ludendorff Bridge
Rennie, Thomas, 246
Reynolds, John, 134
Rheinberg, 99–102
Rhoades, D. F. "Dusty," 261
Rhodes, Charles, 112
Richardson, Walter B., 69, 79, 191
Ridgway, Matthew B., 144, 196–197, 199, 214, 275–276, 278–279

Ritchie, Neil, 223
Roer River dams, 19–20
Rood, Finous L., 261–262
Roosevelt, Franklin D., 5–6, 8, 284, 288
Rose, Maurice, 67, 68–69, 78, 79, 90, 91, 92, 107, 108, 161
Roseborough, Morgan G., 99
Rosenman, Leon, 113
Ross, Cecil F., 38
Rothkirch und Tritt, Ernst Georg Edwin Count von, 162
Royal Air Force, 4–5, 20, 106, 160, 220, 224, 232, 280
Rundstedt, Gerd von, 28, 33, 45, 67, 71, 72, 122, 152, 155
Russian army, 3, 195, 197, 284, 285
Rust, Clayton A., 177
Ryan, Cornelius, 287

Saar-Palatinate, 162–163, 166, 179, 181, 184
Sabia, Carmine, 126, 138
Salisbury, Joe, 266
Saunders, Pat, 200
Scalan, Enoch J., 168
Schaper, Jack F., 75
Scheller, Hans, 121, 125–126, 135, 139, 141, 159
Schellman, Robert H., 29–30, 31
Schimpf, Richard, 151
Schlemm, Alfred, 87, 97–98, 102, 202–203
Schloss Diersfordt, 242, 243
Schloss Rath, 61–63
Schloss Schlenderhan, 94–97
Schreiber, Hans, 178
Schulz, Rudolf, 140, 141
Schwammenauel Dam, 20, 22–23, 27, 28, 29–31, 44
Sculkowski, Simon, 264
2nd (Hell on Wheels) Armored Division, U.S., 66–67, 81, 84, 87–88
Second Army, British, 7, 206
II Corps, Canadian, 20
II Parachute Corps, German, 214, 220
2nd Parachute Division, German, 87
Seventh Army, German, 69–70, 184
Seventh Army, U.S., 7, 163, 166, 170, 171, 181, 183, 184, 185
VII Corps, U.S., 22, 28, 45, 58, 67, 68–69, 75–76, 77, 78, 90, 160

17th (Thunder from Heaven) Airborne Division, U.S., 199, 207, 208–209, 210, 212–213, 214, 227, 228, 229–230, 231, 232, 258, 271, 275–276, 277–278, 280

17th Armored Engineer Battalion, U.S., 88

LXXI Corps, German, 97

78th Infantry Division, U.S., 19, 22, 29, 148, 158

79th (Cross of Lorraine) Infantry Division, U.S., 200, 201–202, 223

771st Tank Battalion, U.S., 85

784th Tank Battalion, U.S., 98

SHAEF (Supreme Headquarters, Allied Expeditionary Forces), 7, 13, 14, 18, 25, 29, 32, 44, 83, 173, 179, 195, 273, 274, 284

Sheppard, Burleigh, 43

Shuit, Lawrence J., 91

Siegman, Harold, 112

Siergiej, Edward J., 276–277

Sigerfoos, Edward, 276, 277

Simonds, Guy G., 20

Simpson, William H., 11, 12, 17, 19, 27–28, 44–45, 53, 58, 66, 83–84, 89–90, 200, 274, 281, 287–288

Sinzig, 128

6th (Red Devils) Airborne Division, British, 199, 207, 208, 212–213, 214, 227, 229, 230–231, 243, 245, 250, 257, 258, 267–268, 269, 271, 277–278, 280

6th Army Group, U.S., 7

643rd Tank Destroyer Battalion, U.S., 84

Skorzeny, Otto, 153, 178

Slate, Max L., 75

sleeper patrols, 38–42

Smith, Turman, 117

Smith, Walter B. "Beetle," 7–8

Smythe, George W., 148

Sobert, Willard, 222

Spaatz, Carl, 287

Spaulding, D. L., 186–187, 189

Speer, Albert, 34, 204

Stalin, Joseph, 5, 6, 197, 284, 285, 287

Steeg, 72–73

Stelling, William, 42–43

Stelljes, Karl H., 38–39

Stiller, Alexander, 225, 226

Straube, Hans, 243

Ströbel, Hans, 159

Stryker, Stuart S., 256

Summers, William M., 54, 55

Summersby, Kay, 12

Sweat, Wesley A., 91–92

Taylor, Fred, 220

Taylor, Maxwell D., 144, 196

3rd Armored Division, U.S., 67, 68–69, 78, 79–80, 90, 97, 105, 106–107, 108, 116, 161

Third Army, U.S., 7, 11, 32–33, 143, 152, 161, 162–163, 166, 179, 185, 190, 224, 283, 284

XIII Corps, German, 33

XXX Corps, British, 20, 26, 32, 34, 66, 219–220, 245

30th (Old Hickory) Infantry Division, U.S., 45, 53, 200, 201–202, 221–223

35th (Santa Fe) Infantry Division, U.S., 63–64, 98

Thompson, C. R. "Tommy," 204

Thompson, John H. "Beaver," 114–115

Thomson, Robert, 266–267

303rd Engineer Battalion, U.S., 30–31

313th Troop Carrier Group, U.S., 255

320th Infantry Regiment, U.S., 63–64

334th Infantry Regiment, U.S., 75

379th Infantry Regiment, U.S., 87

385th Field Artillery Battalion, U.S., 44

Thurston, Clair, 154

Tieg, Mervin, 108–110

Timmerman, Karl, 123–124, 126, 128, 129, 134, 135, 136–137, 138, 176, 288

Topham, Frederick G., 244–245

Tracha, Eldred, 262–263

Truitt, Howard, 277–278

12th Army Group, U.S., 7, 8, 12, 190, 285

XII Corps, British, 223

XII Corps, U.S., 168, 181

XX Corps, U.S., 166, 168

21st Army Group, British, 12, 22, 180, 195, 198, 200, 201, 272, 273, 274, 283

23rd Armored Engineer Battalion, U.S., 174

28th (Keystone) Infantry Division, U.S., 20, 25

29th (Blue and Gray) Infantry Division, U.S., 45, 53, 82

Ultra, 28–29, 45, 163, 165, 204, 206–207, 226

Valsangiacomo, Oreste V., 103–104
Vandenberg, Hoyt, 224, 287
Van Fleet, James A., 178
Van Houten, John H., 99, 100
Varrone, Lou, 211
Varsity, Operation, 199, 200, 205–216, 227–271, 272, 275, 280–281
Vaughn, Lynn, 253
Vergeltungswaffen (vengeance weapons), 175–176, 230–231
Veritable, Operation, 20, 22, 26–27, 31–32, 45, 66, 67
Vermillion, Robert, 228, 268
Volkssturm (people's army), 3–4, 65, 79, 97, 152, 169, 195

Walker, Calvin, 61–63
Walker, Walton H., 166, 168
Waltz, Welcome P., 54
Waters, Fred D., 60
Waters, Ray, 46
Watkins, James M., 63–64
Weber, Günther, 207
Wehrmacht, 7, 13, 29, 83, 86, 157, 158, 180, 183, 190–191, 205, 207, 218, 267, 286
Weishaupt, Donald R., 39
Welborn, John C., 107

Werewolves, 13–14
Wesel, 202, 220–221
Wesel Bridges, 94–105
Wesel Division, German, 220–221
Westphal, Siegfried, 158
White, Isaac D., 66–67, 81
Whitely, John F., 12, 17–18
Wiggins, Buford, 117
Williams, Adriel N., 259–260, 261
Williams, Arthur E., 36–37
Williams, Charles "Chuck," 265–266
Williams, Paul L., 206
Wolff, Karl, 190–191
Wolfson, Mike, 150–151
Wood, William B., 86
Worrilow, Charles L., 252–253
Wunderwaffen (wonder weapons), 24–25, 59–61
Wyche, Ira, 223

Yalta Conference, 5–6
Yeomans, Prentice E., 92
York, Alvin, 196
Young, Mason J., 173–174
Youngblood, George, 88–89

Zangen, Gustav von, 121, 125
Zegerheim, Emil, 39
Zhukov, Georgi, 197
Zichterman, Jacob, 280

Here's How...Here's Help

HOW TO BUY A CAR by James R. Ross
The essential guide that gives you the edge in buying a
new or used car.
_____ 90198-4 $3.95 U.S.   _____ 90199-2 $4.95 Can.

MARY ELLEN'S HELP YOURSELF DIET PLAN
"The one-and-only good plan...presented by a funny,
practical lady." —*Kirkus Reviews*
_____ 90237-9 $2.95 U.S.   _____ 90238-7 $3.50 Can.

WHEN YOUR CHILD DRIVES YOU CRAZY
by Eda LeShan
Full of warmth and wisdom, this guide offers sensible
advice spiced with humor on how to weather the storms
of parenting.
_____ 90387-1 $4.50 U.S.   _____ 90392-8 $5.75 Can.

THE CHICKEN GOURMET by Ferdie Blackburn
A mouth-watering celebration of over 100 international
and classic recipes for family and festive occasions.
_____ 90088-0 $3.50 U.S.   _____ 90089-9 $4.50 Can.

THE WHOLESALE-BY-MAIL CATALOG—UPDATE 1986
by The Print Project
Everything you need at 30% to 90% off retail prices—by
mail or phone!
_____ 90379-0 $3.95 U.S.   _____ 90380-4 $4.95 Can.

HOW TO GET A MAN TO MAKE A COMMITMENT
by Bonnie Barnes and Tisha Clark
Take charge of your life—discover the two-week plan to
get your relationship just where *you* want it to be!
_____ 90189-5 $3.95 U.S.   _____ 90190-9 $4.95 Can.

*NOW AVAILABLE AT
YOUR BOOKSTORE!*